Warfare and Violence in the Iron Age of Southern France

Mags McCartney

BAR International Series 2403
2012

Published by

Archaeopress
Publishers of British Archaeological Reports
Gordon House
276 Banbury Road
Oxford OX2 7ED
England
bar@archaeopress.com
www.archaeopress.com

BAR S2403

Warfare and Violence in the Iron Age of Southern France

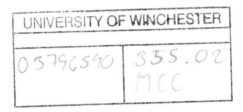
ISBN 978 1 4073 0998 9

Printed in England by Information Press, Oxford

All BAR titles are available from:

Hadrian Books Ltd
122 Banbury Road
Oxford
OX2 7BP
England
www.hadrianbooks.co.uk

The current BAR catalogue with details of all titles in print, prices and means of payment is available free
from Hadrian Books or may be downloaded from www.archaeopress.com

PREFACE

Much of the following research is taken from a doctoral thesis of the same name, submitted at Queen's University, Belfast in 2007. Since then I have continued to develop many of the themes explored in this monograph in a series of forthcoming papers. Nevertheless, the case study presented here still represents a new approach to identifying and interpreting prehistoric warfare. The archaeological material too, remains little-known in English-speaking archaeology. It is hoped that, in particular, the presentation of chronological syntheses of the major excavated sites, iconography and human remains, will be useful; even to those who reject my conclusions on their significance within the societies which created them.

Mags McCartney
Edinburgh, January 2012

ACKNOWLEDGEMENTS

I would like to thank Ian Armit, University of Bradford; Mark Gardiner, Queen's University Belfast; Rick Schulting, University of Oxford and Ian Ralston, University of Edinburgh for their guidance and support. Thanks also to Xavier Delestre and all the staff at the SRA in Aix-en-Provence for allowing me access to the excavation records for Provence. In addition various archaeologists working in the region, notably Patrice Arcelin, Thierry Janin, Kevin Walsh and Loup Bernard, were very helpful in answering questions regarding specific aspects of the archaeology and general approaches to the material in France. Carole and Melvin Ember from the Human Resources Area Files at Yale gave advice and support throughout this study, as well as encouragement regarding potential future work in this field. I would also like to thank the many friends and colleagues who gave me technical advice. Chris Knüsel, University of Exeter and Linda Fibiger, University of Edinburgh in particular both offered invaluable advice on the osteological remains and Linda commented on the chapter dealing with this. The photograph of the Iban trophy heads was taken by Janet Bell while doing voluntary work in Southeast Asia. Much of the text was written by Ms P.Z. McCartney (this was all subsequently deleted but her contribution was greatly appreciated). While this study would have been impossible without much guidance and advice, the research and its conclusions, including all errors and omissions, are the author's.

ABSTRACT

The Iron Age of southern France remains relatively unknown in the English-speaking archaeological world. The best-known aspects of the archaeological material suggest a society in which warfare was an overriding preoccupation. Major, fortified centres, such as Entremont and Saint-Blaise, and the tradition of 'warrior statues' like those from Entremont and Roquepertuse, suggest that conflict was a recurrent theme. Literary sources, such as Poseidonius have described the indigenous populations of this area as a volatile and warlike people who took the heads of their enemies from the battlefield and displayed or preserved them in their settlements. Finds of skulls, some with nails still embedded in the bone, appear to verify such reports. Notwithstanding the recurring theme of conflict, these strands of evidence have never been analysed in tandem to identify the role and representation of warfare during this crucial period of colonial contact between the indigenous populations of the region and the state-level societies of the Mediterranean.

Warfare is a relatively new discipline in archaeological study and it has been suggested that previously archaeologists have been guilty of 'pacifying the past' (Keeley 1996). While recently archaeologists have proved more willing to identify indications of past violence, the tendency to treat warfare as a series of isolated, episodic events persists. This is in stark, contradistinction to warfare studies in social-anthropology, which view conflict as an unfortunate but 'normal' form of social interaction. This study aims to identify patterns of warfare in the southern French Iron Age through examination of the documentary, settlement, iconographic and osteological evidence for warfare in this region, each within its chronological context and in tandem with one another. The pattern of warfare which emerges from this analysis will then be discussed within some of the more prominent models of social-anthropological study. This case-study offers a more nuanced and contextual interpretation of warfare in the southern French Iron Age and demonstrates how, if treated as a form of social interaction, rather than a breakdown in social norms, might be integrated into wider archaeological interpretations of social and political change.

TABLE OF CONTENTS

LIST OF FIGURES

CHAPTER 1 INTRODUCTION

1.1 General Introduction

Situated around the Rhône Corridor, the 'Celto-Ligurian' region of southern France formed a bridge between the Mediterranean world and the northern societies of Iron Age Europe. This cultural frontier-zone has produced a range of archaeological evidence which appears to relate to the theme of warfare. These include fortified, nucleated settlement, statuary representing warriors, and human remains which suggest an apparently war-related head-cult. Despite a dynamic and engaging tradition of archaeological research in southern France, much of this material remains relatively unknown in the English-speaking archaeological world. Notwithstanding the recurring theme of conflict, these strands of evidence have never been analysed in tandem to discern the nature and extent of warfare in the region during the crucial period of colonial contact between the indigenous cultures of the region and the state-level societies of the Mediterranean.

1.2 Aims of the Study

This study will examine the nature and extent of warfare in the Iron Age of southern France from the end of the seventh century BC until the early-first century BC. This comprises the period from the emergence of enclosed, nucleated settlement among the indigenous populations and the first permanent Mediterranean presence in Provence, until the annexation of the province by Rome and the end of traditional indigenous settlement and cultural expression in the region. The evidence for warfare will then be considered in relation to wider theoretical interpretations of the causes and development of warfare in non-state societies. A range of data will be examined in turn, including literary sources, settlement structure and organisation, iconographic representations of warriors and warfare, and the osteological remains of assemblages from key sites. The disparate strands of evidence will be brought together in an attempt to ascertain patterns of defensive or offensive activities throughout the period. These patterns will then be evaluated within a theoretical model to elucidate the degree of conflict and its social role in southern France during the Iron Age.

1.3 Warfare and Society

Violence is common to all human societies. War is not (Ferguson 1984, 12) but its occurrence is so frequent and widespread as to provoke the question why some few societies do not resort to warfare (Haas 1990, xii). At what degree of organisation or scale of numbers violence becomes war is still the subject of some debate (e.g. Mead 1968; Ferguson 1984; Mercer 1999). What is certain, however, is that warfare has been a preoccupation of the earliest written records of the Egyptians, Chinese, Greeks and Romans, and is a dominating theme of most of the oral histories and mythologies of preliterate societies (Keeley 1996, 3-4). Since the Enlightenment war has been used to epitomise conflicting definitions of the human condition, and it is the language of the Enlightenment which is still used to describe differing interpretations of the causes and mechanics of war in social anthropological studies. For example, the term 'Rousseauian' (ibid., 20) has been used to reflect interpretations which view warfare as the result of ecological, political or other influences, in reference to Rousseau's proposal that the pre-civilised or 'natural' condition of humans was one of peaceful co-existence. The term 'Hobbesean' (Keeley 1996, 16) has been used to describe interpretations which see peace as an artificial situation, the maintenance of which requires overarching political interference. This is in allusion to Hobbes's picture of non-state societies being involved in an almost continual condition of conflict.

Social anthropology has produced a range of interpretations to explain the origins of warfare, the main themes of which, materialist, biological and cultural, are discussed more fully in the following chapter. Ethnographic and ethno-historical accounts are, necessarily, snapshots from societies experiencing some degree of contact with state-level societies. Observations from social anthropology lack a long sequence of events in which to study, for example, cycles of war and peace, how these change before and after contact with state-level societies, and how they relate to cultural expressions of warfare. The comprehensive study of patterns of warfare requires the time depth of the archaeological record (Haas 1990, 175). Ethnographic studies also come with subjective assertions from participants regarding motivations and causes (Keeley 1996, 117; Ferguson 1984, 41). For this reason Keeley suggests that the 'silent' evidence of archaeology might give a less ambiguous picture of demographic, ecological or ideological factors (1996, 117).

Lately, archaeology has been employed to study patterns of warfare in the American southwest (Leblanc 1999; Lekson 2002). In Europe the explicit study of prehistoric warfare has emerged only recently, in part inspired by Keeley's publication *War Before Civilisation* (1996). Subsequently, a range of European-based publications began to explore facets of prehistoric archaeology in relation to possible models of conflict. *Material Harm* edited by John Carman (1997) examines a variety of approaches to violence and warfare in the archaeological record worldwide. *Troubled Times* (1997) edited by Martin and Frayer, draws on archaeological, ethnographic and osteological evidence for warfare in a variety of societies. *Warfare in the Late Bronze Age of Northern Europe* (Osgood 1998) observes a wide variety of evidence sources which imply warfare in a contained area and period (1998). *Ancient Warfare* edited by John Carman and Anthony Harding (1999) broaches several aspects of warfare studies from a variety of time periods; and *Bronze Age Warfare* (Osgood, Monks and Toms 2000) focuses on the evidence from a variety of Bronze Age sites throughout Europe collating a range of archaeological data, including artefacts, iconography and settlement evidence.

Western Europe offers the opportunity of researching patterns of warfare through one of the widest and most extensively studied archaeological records in the world. To date, however, studies of Iron Age warfare in Europe have mostly consisted of sporadic considerations of specific and disparate topics. Most studies are very limited in the scope of the evidence considered, being restricted, for example, to typologies of particular artefacts (e.g. Bridgforth 1997) or structures (e.g. Avery 1986; Bowden and McOmish 1987, 1989). Questions of political and social organisation and how these relate to cycles of warfare in Bronze Age and Iron Age Europe, however, have rarely been investigated, notable exceptions being Kristiansen (1999) and Randsborg (1999).

This study will examine the extensive body of evidence from southern France during the Iron Age, in particular the area around the Lower Rhône Valley. This region witnessed the emergence of permanent, nucleated, indigenous settlement, and has been called the 'crucible' of urbanisation in southern France (Chausserie-Laprée 2005, 58), as well as being the location of Massalia, the first permanent Greek colony in southern France (Py 1993, 90). The Lower Rhône Valley is also strongly associated with many of the archaeological remains which appear to reflect a society preoccupied by warfare. Representations of warriors have been discovered in Languedoc, however, the major tradition of 'warrior statues' is concentrated around the Lower Rhône Valley. Again, while skulls and fragments of skulls have been frequently discovered in Languedocien settlement sites, four of the five sites associated with the display or deposition of complete or near-complete crania, are found in Provence. This region reflects many of the issues brought into prominence from social anthropology, for example, the effects of contact between societies of different socio-political complexity, as well as, apparently, displaying a concentration of many aspects of the enculturation of certain warlike values which appear to occur throughout Europe during the Iron Age.

1.4 The Re-evaluation of 'Primitive' Warfare

The neglect of warfare in archaeological study is, perhaps, partly to do with an over-simplification of perceptions of past societies. Keeley believes that 'primitive' or tribal warfare has been considered so ineffectual as to be harmless, and as such has been relegated to the sphere of ritual activity (1996, 9-11). Warfare studies may also have been hampered by the elusive or subjective character of much of the evidence. For example, many weapons may also have been tools and vice versa (Mercer 1999, 143; Bridgforth 1997, 95), and the defensive intention of certain structures can be debatable (Mercer 1999, 143; Avery 1986; Bowden and McOmish 1987, 1989). However, building on the models devised within the more developed warfare studies of social anthropology, European archaeology is in a position to test many of these hypotheses by drawing on a

database of extraordinary time-depth and rigorously studied strands of evidence.

The current emergence of interest in warfare studies in archaeology has been variously explained as a reaction to the many ethnic wars of the 1990s (Vandkilde 2003, 126), or the urgency at the end of the twentieth century to understand the global conflicts of that period (Gilchrist 2003, 1). A major catalyst has undoubtedly been Keeley's *War Before Civilization* (1996). His criticism of the 'pacification of the past' in the literature of archaeology and anthropology has precipitated a torrent of publications. Keeley argues that archaeological evidence and ethnographic testimony have been distorted in two different, but equally misleading ways. Firstly he believes that since the devastation of the Second World War, archaeologists and anthropologists have viewed prehistory and 'primitive' or non-state societies, as existing in a peaceful golden age. Evidence for war outside 'civilisation' is interpreted as ritualised and relatively harmless (Keeley 1996, 22); weapons, for example, being viewed as status symbols, rather than military equipment (*ibid.*, 19) and actual violence being seen as a last resort in situations of great need. Keeley sees the principal application of this premise in materialist models which view warfare as too expensive and risky an undertaking to be considered unless absolutely necessary. In reference to Rousseau's assertion that 'man is an enemy to others only when he is hungry' (*ibid.*, 15), Keeley calls this model 'neo-Rousseauian' (1996, 20). Secondly, Keeley decries the interpretation that 'primitive' or ancient societies were prone to constant outbreaks of conflict due to the lack of centralised control, but that this same disorganisation resulted in ineffectual warfare. In an allusion to Hobbes's premise that the natural human state was one of conflict rather than peace, Keeley dubs this position 'Neo-Hobbesian'. Although approaching the subject of prehistoric warfare from very different positions, Keeley charges both of these with misrepresenting non-state level war. Neo-Hobbesian war may be the eternal social condition, but, without the tactical or technical superiority of 'civilised' warfare, it is inconsequential. Neo-Rousseauian war is rarely employed and only out of necessity. In either case, Keeley feels they perpetuate the myth of 'primitive warfare' (*ibid.*, 16-17).

One of the most provocative aspects of Keeley's study is his assessment of ethnographic examples of war by regarding war-casualties in relation to the numbers of those involved in combat or the overall demographic of the participating society (1996, 88-94). This approach is not entirely new, as Gardner and Heider (1974, 144), in their record of the notoriously warlike Dani tribe of Papua New Guinea reckoned that Dani losses in all kinds of combat amounted to not more than those dying of colds and other illnesses each year. Cohen (1990, 330) also, specifically refutes the idea that non-state warfare is 'less lethal' by comparing the relatively small percentage of adult male fatalities in war from Europe and the United States with a much larger proportion of the Murngin tribe of north Australia. However, Keeley uses this re-evaluation of the destructive consequences of 'primitive' warfare to highlight the potentially overwhelming impact of conflict in non-state level societies.

1.5 Definitions of Warfare

This examination of the scale and effects of conflict in relation to the participants prompts the reassessment of 'warfare'. Mercer points out that a variety of aggressive behaviours from actual raiding to display activities are carried out which are too wide-ranging and ambiguous to be universally defined as war (1999, 144), so when does violence become warfare? – and can this help us to understand why it happens?

Degrees of scale in conflict do not fit neat definitions of warfare, as small-scale feuding may either precipitate or be part of larger violent events (Ferguson 1984, 4; Keeley 1996, 44). Scale, as discussed above, when viewed as a percentage of the social unit involved in the conflict, varies with each event (Keeley 1996, 64, 88-94). Likewise classifications depending on the mobilisation of autonomous groups are problematic as definitions of autonomy are fluid (McCauley 1990, 2), and Ferguson cites examples such as revolutions and civil wars where alliances unite and hostilities divide only for the duration of the conflict (1984, 4). In their statistical analysis of predictors of warfare Ember and Ember (1992, 248) defined warfare as 'socially organised armed combat between members of different territorial units (communities or aggregates of communities)'. Most other definitions are variations on this theme, such as "Mead's (1968) 'groups...in purposeful, organised and socially sanctioned combat involved in killing'" (McCauley 1990, 2). While each definition differs, these are differences of degree and are not necessarily contradictive. The range of variation in each occurrence of warfare works against anything other than broad and flexible definitions, however, the received interpretation is that warfare must be examined in terms of social policy, as a socially sanctioned event (as opposed to, for example, murder); and on a polity level, whatever the size or duration of the political unit (Mercer 1999, 144; Ferguson 1984, 5; McCauley 1990, 1-2). Generally classifications are kept deliberately broad, such as Ferguson's definition of war as 'organised purposeful group action directed against another group, that may or may not be organised for similar action, involving the actual or potential application of lethal force' (1984, 3-5). Sharples (1993) points out that this definition is a particularly useful one to apply to archaeological studies. For example it covers situations such as 'cold wars' which take place without fatalities, only the threat of killing. It also uses the term 'organised' rather than 'political' groups, omitting the necessity of identifying such specific socio-political units in the archaeological record (Sharples 1993, 80). This, therefore, is the definition which will be adhered to here.

1.6 Structure

Chapter three outlines a general archaeological background of the south of France from the late-seventh century until the first century BC. After this, the main strands of evidence as laid out in the introduction will be examined. The documentary evidence from Greek and Roman sources relating to the period is discussed in chapter four. Chapter five examines evidence from settlement patterns and the construction of oppida in relation to defensive strategies throughout the period. Chapter six considers iconographic evidence for warfare, in particular the warrior statues and engraved and sculpted representations of severed heads found in sites of the Bouches du Rhône, such as Entremont and Roquepertuse, and how these may represent warrior ideologies. Chapter seven reviews the osteological evidence for warfare, including, where possible, palaeopathelogical examination of human remains, and an assessment of the cranial remains which were displayed as part of an apparently pan-European 'head cult' which is in evidence in the skulls at the Bouches du Rhône sanctuary sites.

Chapter eight discusses how the disparate strands of evidence are brought together against the archaeological and documentary chronologies to attempt to ascertain patterns of warfare throughout the period. Finally, the implications of these diverse strands of evidence are then studied in relation to known correlates of war. Social anthropology offers a wide-ranging record of the various contexts of war: how it is carried out, and the consequent rewards and drawbacks to political, social and economic structures. For example, war as a mechanism to develop increased political complexity (Carneiro 1970; Earle 1997), or the destabilising effect of recurrent warfare on society (Ember and Ember 1992). The following chapter, therefore, covers the current theories on the causes of warfare as discussed in social anthropology.

2.1 Introduction

The traditional approach to tribal warfare, in both anthropology and archaeology, has been cultural, in as much as non-state societies have been regarded as engaging in conflict for revenge or to gain prestige rather than for any polity-level strategy for gaining territory or resources (Ferguson 1984, 23). Recent anthropological study has attempted more nuanced approaches to warfare. In addition to cultural motivations for warfare, these new approaches take account of material motivations, which explain warfare as a means to gain or keep resources, and biological imperatives, which see warfare as a social expression of innate human aggression. There is a certain amount of overlap within these classifications and most anthropologists regard their work as a more specified subset of one or more of these approaches. For example, *Warfare, Culture and Environment* (Ferguson 1984) is a collection of various materialist subsets, 'cultural materialist' approaches (Harris), 'ecological and cultural factors' (Biolsi) and materialist motivations in relation to socio-political factors (Cohen). *The Anthropology of War* (Haas, 1990) uses slightly different headings,' ecological-materialist', 'bio-cultural' and 'historical'. A more recent paper by Ferguson (2001) reduces the current interpretations of warfare to three overarching approaches, 'biological', 'cultural' and 'materialist'. These three interpretations effectively cover the majority of hypotheses and for present purposes it is useful to follow this classification.

The idea that violence is an innate human response does not in itself explain the occurrence of warfare, which is essentially a social act. Biological theories, have, however, been used to explain the widespread evidence for human aggression, and socio-biological theories have been employed to account for the development of cultural institutions which support organised acts of group violence. A biological predisposition for violence is repeatedly cited as a ubiquitous and universal explanation for warfare (Ferguson 1984, 8), although usually as an ancillary aspect.

Materialists tend to view their interpretation not only as the primary cause of war in itself but also as an underlying factor in other interpretations. For example, if the stated impetus for war is revenge (i.e. cultural), the origins of this motivation will nevertheless often be considered to have a utilitarian reason (i.e. materialist) behind it. The materialist interpretation is proposed as an all-pervading and universal factor in warfare, therefore the validity of this approach needs to be discussed on its own to argue its merits.

This leaves three main issues to be discussed, cultural interpretations, which perceive warfare as an integral part of certain social structures, historical interpretations, which view each event specific to its own circumstances as influenced by individual participants, and socio-

political, also referred to as functionalist, interpretations, which view warfare as being controlled by political structure. All three of these approaches can be seen as attempts to take account of the distinct and individual nature of particular conflicts. They are also all usually discussed as aspects of multi-factorial interpretations. Therefore, while it might be useful to follow Ferguson's recent categorisations of biological, materialist and cultural, 'cultural' interpretation might best be regarded as an umbrella term for the individualistic nature of warfare as a reflection of a society's ideology, socio-political complexity and response to specific events.

2.2 Biological Interpretations

Origins and Definition

Biological interpretations of warfare centre on the theory that aggression is an innate human characteristic. This idea is believed to have been formalised after the First World War with Freud's conception of human aggression as an innate instinct for self-destruction which he called the 'death-drive' (Ferguson 1984, 9). This hypothesis remained unexplored in social anthropological studies until the late 1960s, when Lorenz published *On Aggression*, an examination of warfare as the expression of innate human aggression (Layton and Barton 2001, 13; Lorenz 1978). Lorenz, in his 'drive-discharge' theory proposed that there was an innate human drive for aggression which built up until it found a stimulus which caused it to discharge (Ferguson 1984, 9). The stimulus would result in acts of aggression within a communal, cultural context (e.g. warfare), but the violence triggered could be seen as a strong phylogenetic trait (Lorenz 1978, 70). Lorenz suggested that humans undergo a phase once during the development of conceptual thought, when they adopt new and very strong 'object fixation' a procedure known in animal behaviour as 'imprinting' (*ibid.*, 69). This object fixation, which can take the form of an ideal or a political affiliation, was responsible for forming the stimulus situation which initiated the aggressive instinct in his 'drive-discharge' theory. According to this model, the stimulus overrides 'instinctive inhibitions' against killing. "Men may enjoy the feeling of absolute righteousness even while they commit atrocities...As the Ukrainian proverb says: 'when the banner is unfurled, all reason is in the trumpet'" (*ibid.*, 69-70).

Lorenz's *On Aggression* (1978) had much the same effect on social-anthropology, as Keeley's *War Before Civilisation* (1996) has had on archaeological warfare studies. The idea of aggression as a result of natural selection was instantly popularised and subsequent publications were informed by growing interest in ethology and psychology (Ferguson 1984, 9). Comparative studies of human and animal aggressive behaviours were made. Humans were distinguished from other animals by the development of culture. Human behaviour, therefore, was divided into genetic evolution and cultural evolution or social biology (Tinbergen 1978). For proponents of social biology, culture or 'intellectual emancipation' (Wynne-Edwards 1978, 104) can be seen as another aspect of evolutionary development. Humans, in other words, have culture because they evolved biologically to develop culture. Tiger and Fox (1978) propose that short of 'divine intervention' it can only be supposed that human behavioural predispositions, including the ability to create culture, developed in the same way as other animals: 'by mutation and natural selection' (*ibid.*, 57).

Warfare and Human Evolution

In an attempt to test the theory that humans have been shaped by their ancestral environment to fight for territory, reproduction opportunities and status (Thorpe 2003, 147), Layton and Barton conducted a large cross-cultural study using data from a range of ethnographic studies. They began by considering a study regarding chimpanzee aggressive territorial behaviour. Studies comparing chimpanzee and human behaviour had traditionally been put forward to bolster aspects of Lorenz's theory that attribute warfare to instinctive territorial defence. This ethological research proposed a number of factors.

The first hypothesis of the study was that chimpanzees and humans are similar to each other and atypical of other animals in that they kill members of their own species. This has been used to suggest that warfare may have been an ancestral characteristic before evolutionary divergence (Layton and Barton 2001, 13). However, Layton and Barton consider the case for chimpanzee inter-group killing to be over-stated as there is only one definitely known example of a chimpanzee group being destroyed in inter-group raiding (*ibid.*, 15). They also assert that the chimpanzee groups used in the study could not be considered 'species typical' (*ibid.*, 14) due to the possible effects of human interference (Layton and Barton 2001, 14-15). Thorpe argues that comparisons between chimpanzee and human aggressive behaviour do not correlate well, the core idea of territoriality being that the aggressive territorial instinct among chimpanzee groups is driven by a distrust of strangers, whereas many accounts of 'primitive' conflict describe fighting as taking place between groups which are familiar with each other such as Yąnomami tribes in the Amazon and Dani tribes in Papua New Guinea (Chagnon 1996, Gardner and Heider 1974).

A second hypothesis of chimpanzee-human comparative studies proposes that chimpanzees and small human hunter-gatherer groups resemble each other in that both males and females will join their partner's group. Among most other primates, only males leave their group at puberty to reproduce. This has been used to extrapolate an evolutionary interpretation of human inter-group alliances (Layton and Barton 2001, 14). Layton and Barton argue that the patrilineal and/or patrilocal practices of some hunter-gatherer groups are fundamentally different from chimpanzee organisation, in that the former is ideological and the latter is behavioural (*ibid.*, 15).

Thirdly comparative research had suggested a further commonality between human hunter-gatherer groups and chimpanzees, in that both will employ boundary defence and cross-boundary raids when prompted by population densities. However, Layton and Barton argue that the practices are fundamentally different, in that human groups employ a concept of regional-level territory and access of passage among allies, unknown among chimpanzees, and this more flexible form of territoriality is sustained and manipulated through a series of reciprocal relationships (*ibid.*, 19-22).

A major consideration of evolutionary determinism proposed by Layton and Barton is that, as small-scale human societies occupy many different habitats and pursue very different subsistence strategies, it is necessary to look at the influence of ecology and subsistence on territoriality. To this end they compiled a wide-ranging set of data from anthropologists recording the territorial behaviours of a variety of hunter-gatherer groups in a variety of ecological conditions and dependant on a variety of different subsistence strategies. Two sets of data were particularly relevant to the study. Firstly those living in low-latitude rain forest environments analogous to the chimpanzee ecological niche (e.g. the Mbuti of the Ituri forest and the Batek De of Malaysia). Secondly those from semi-arid environments like those in which modern humans are thought to have evolved (e.g. the Dobe !Kung of the Kalahari and the Yankunytjatjara of the Australian Western Desert) (2001, 18-9).

Referring to the rain forest-dwelling Yąnomami, Layton and Barton suggest that, while members of that society deny that violence was employed for territorial reasons, they were known to have pushed out neighbouring tribes, so that there was an aspect of territoriality in their warfare (Layton and Barton 2001, 19). However, it was also established that the boundary defence seen in chimpanzee territorial behaviour was not universal in human groups and while humans established boundaries, they remained flexible regarding rights of 'ownership' (*ibid.*, 18). The difference between chimpanzee and human maintenance of territorial boundaries was particularly marked in the ecological niche associated with human evolution. These arid environments offered less dependable resources than the rainforest and the constant threat of drought was universal to all human groups living in these places. In this situation, rather than engendering defensive or hostile territoriality, various bands consistently employed a system of mutual access to each other's territories (Layton and Barton 2001, 19). This co-operative behaviour and its contrast to aggressive chimpanzee territoriality led Layton and Barton to conclude that human territorial strategies are matters of ideology rather than instinct, and developed after the two species diverged (*ibid.*).

Conclusion

In general the biological theory has not been given much credence in archaeological attempts to explain the occurrence of warfare. The propensity for variation in human responses in different situations refutes the simplistic nature of biological theory (Stewart and Strathern 2002, 10; Eisenberg 1978, 172-73). Wilson notes that historically and at all stages of social organisation, warfare and territorial expansion are closely connected, an association which he believes indicates a genetic predilection for war on account of its generating genetic success (1978, 295). For others, the relationship between innate behaviour and learned responses is more complex (Tinbergen 1978, 92; Irons 1979, 9; Alexander 1979, 65). In answer to Wilson in particular and socio-biologists in general, Gould makes a distinction between biological potential and biological determinism. He argues that while human evolution may have genetically programmed humans for aggression, the same biological processes allow peaceful behaviour. A variety of disparate biological predispositions are determined by social structures in a variety of disparate cultures. The human brain is 'capable of the full range of human behaviours and predisposed towards none' (1978, 349).

2.3 Materialist Interpretations

Origins and Definition

In the 1940s there was a brief period of interest in the economics of war in anthropology when the historian Hunt (1940) suggested that Iroquois wars were fought to control the trade in furs (Ferguson 1984, 23). Prior to this, motivations of prestige or revenge were thought to be more important in non-state level societies. After the Second World War materialist theory focused on ecological or environmental explanations of warfare (*ibid.*, 25). By the 1960s ethological research, which examined animal behaviour, began to influence the debate. From these ethological studies derived the branch of research known as cultural ecology, which studied how human societies adapted to and sustained themselves in their environment (*ibid.*, 28). This initiated the beginning of a specific study of warfare in anthropology as one of a variety of possible adaptations to ecological stress. From the 1970s and 1980s this school of thought was reappraised and currently the ecological interpretation of war focuses, not on pure ecology (population densities, carrying capacities, and so on), but on the relationship between the natural environment and the way in which subsistence and social organisation is ordered (Ferguson 1984, 36-37). These can be viewed as mostly internal factors.

Resource Unpredictability, Mistrust and War

It may be difficult to assess material deficiency in a society if, as Ferguson claims, material scarcity is 'relative' (1984, 38), and different levels of political organisation result in different responses to these perceptions of deprivation. Ember and Ember developed a statistical strategy to quantify some of the motivations of people faced with resource paucity. Their statistical analysis used ethnographic data from 186 mostly non-state societies. In quantifying episodes of resource

scarcity, Ember and Ember made the distinction between 'chronic disasters', ecological disasters affecting food-getting resources which can be predicted and 'threat of disasters', that is, unexpected natural disasters which cannot be prepared for (1992, 256). Their results indicated that it was the threat of unexpected, immanent disaster, such as pest infestations or extreme weather fluctuations, rather than current or 'chronic' shortages occurring regularly (*ibid.*, 249), which precipitated war. They found that in virtually all societies where there was a threat of natural disasters, warfare was carried out almost incessantly; and that in these cases the victors usually appropriated resources regardless of whether they were suffering any shortages or not (Ember and Ember, 1992, 256-57). This pattern was particularly apparent in societies with less complex political structures (*ibid.*, 249).

Both of Ember and Ember's major findings were linked to fear, rather than actual or present shortage or threat: 'fear of nature and fear of others' (1992, 257). In an attempt to test Ember and Ember's findings, Lekson applied them to the long archaeological sequence of cyclical warfare in the American southwest assembled by Leblanc (Lekson 2002). Lekson assessed the importance of climatic stress on the agriculture of the region over the period 0-1500 AD and checked Leblanc's construal of periods of ecological stress against dendroclimatological evidence. Lekson believed that annual rainfall in the Colorado Plateau would probably have been a major factor in agricultural success (*ibid.*, 611). He found that each of Leblanc's periods of intensification of warfare was linked to a period, not of climatic extremes, which may have impacted negatively on agriculture, but to periods of climatic unpredictability (*ibid.*, 613), unpredictability being the strongest indicator of warfare in Ember and Ember's analysis.

Warfare in the Civilised-Tribal Zone

In ethnographic case studies Western contact has repeatedly been shown to instigate a destabilizing influence on indigenous economies, social organisation, politics and forms of warfare (e.g. Bennett Ross 1984, Dalton 1977, Beckerman and Lizarralde 1995, Blick 1988). The effects of state-level intrusion on societies with lower levels of socio-political organisation, and the kinds of conflict this can precipitate, are varied and not always linked to direct, aggressive conquest by an expanding state. Even the presence of what Ferguson and Whitehead call 'state agents' (1992, 11) (representatives of state administration such as missionaries, soldiers, government agents or anthropologists) are widely documented as disruptive to the internal relationships of indigenous groups (e.g. Fitzhugh 1985; Gunawardana 1992; Hulme 1986; Law 1992; Washburn 1988).

One of the most striking examples of changes in the patterns and intensification of indigenous warfare brought about by state-level intrusion is found in Ferguson's (1992) anthropological assessment of Western impacts on the Yąnomami. These tribal populations living around the confluence of the Orinoco and Mavaca Rivers of the Venezuelan rain forest are best known through the ethnographic accounts of Napoleon Chagnon, who stayed with the Bisaasi-teri Yąnomami between 1964 and 1966. Chagnon's book *Yąnomamö: The Fierce People* (e.g. 1977, 1983) describes a population for whom inter-group and interpersonal violence was endemic, and who encouraged and rewarded personal attributes of aggression and volatility among men (*ibid.*, 1970). Chagnon has emphasised the role of internal, cultural factors in precipitating warfare among the Yąnomami, claiming that their social systems were near 'pristine', uncontaminated by outside state-level influences (Chagnon 1983, 30, 213-14; Ferguson 1992, 200; Spindler and Spindler 1983, vii). While conceding that, by the standards of many social anthropological case-studies, the majority of Yąnomami groups had, indeed, only limited contact with outsiders, Ferguson claims that much of the violence described by Chagnon can be attributed to outside contact (Ferguson 1992, 201). Ferguson takes account of disasters caused by the presence of outsiders, such as the introduction of new diseases or ecological depletion; and relates these destabilising factors to changes in native infrastructure, again brought about by outside agents.

Changes to Yąnomami Infrastructure

In 1950 the first mission was established on upper Orinoco, followed soon after by other missionaries and settlers from outside. New steel tools had begun to filter into the area in the late 1940s, however, the mission provided a steady supply of steel knives, axes and machetes. Those near the missions had good access to these goods and became wealthy. Groups in the 'interior' or upstream from the supply centres began to harass those in possession of these new, valuable items. In 1958 the government Malaria Control Station was set up at the mouth of the Mavaca River. The central mission-post moved its headquarters to the vicinity, and officials from the Station invited the Bisaasi-teri (the group with which Chagnon was based while conducting his fieldwork) to settle nearby also. The Control Station gained a willing work-force and access to cultivated and hunted food supplies. The Bisaasi-teri meanwhile, in response to the wealth generated by permanent access to valuable steel trade items, formed the first permanent, sedentary Yąnomami settlement in the area (Ferguson 1992, 202-203).

Direct Impacts on Yąnomami Society

The most immediate result of outside contact on the Yąnomami was the great technological advantages in cultivation and construction provided by steel tools. The old trade systems, based on complex networks of marital and political alliances broke down (Ferguson 1992, 214). Along with the new tools came new diseases (hence the foundation of the Malaria Station eight years after the first mission was founded). Malaria and measles had begun to effect populations around the Orinoco-Mavaca since the 1940s, and major outbreaks of malaria have

been noted for 1960 and 1963 (Ferguson 1992, 203). Estimates of the numbers of indigenous peoples killed during such outbreaks vary from between 10% and 40% of the populations (*ibid.*, 203-204). Among the Yąnomami these epidemics were interpreted as witchcraft, perpetrated by neighbouring groups. Around this time, Chagnon (based near the government Station among the Bisaasi-teri) estimated that among the Yąnomami groups in his research area and their allies, 25-30% of adult males were killed in wars (Chagnon 1983, 79), significantly more than reported in other areas during the same period (e.g. Cocco 1972, 393; Lizot 1989, 30; Ferguson 1992, 204). As a result of the Malaria outbreaks and subsequent warfare, even larger numbers are believed to have died, mostly those who had lost providers Ferguson (*ibid.*).

Concurrent with these disasters, depletion of hunt animals was recorded during the contact period (Lizot 1976, 13; Chagnon 1977, 148). The effects of this scarcity brought to light a further impact on Yąnomami social systems. The populations further from the government and missionary centres at the mouth of the Mavaca were not sedentary, and it was common practice among these groups to relocate when resources became depleted or neighbouring tribes became threatening. Among these groups, a system of 'meat sharing' among villages was followed. This practice was enforced by custom and strongly adhered to. In the case of the Bisaasi-teri, relocation was no longer considered an option, and it was noted that, among this group, meat was kept and eaten by the individual family (Chagnon 1977, 91-92; Ferguson 1992, 206).

Ultimate Impact on Yąnomami Society

Among the Yąnomami the direct impacts of state contact (disease and game depletion) were aggravated by the indirect impacts which caused the breakdown of social systems (traditional alliances which had controlled trade networks, policies of relocation and resource-sharing). Ferguson considers the most significant factor in this instability to be what he calls the 'anchoring' (Ferguson 1992, 208) effect. Relocation, a fundamental response to many political and ecological stresses in Amazonia, had a limiting effect on warfare among the pre-contact Yąnomami, and among groups living beyond the sphere of state centres. A further outcome of this instability was the effect on belief systems and attitudes to violence. The myth recorded by Chagnon, that Yąnomami aggression was caused by the blood dripping from a wounded moon, was reported only among another more northerly group (which also experienced recurrent warfare). Among other populations this myth was unknown (Ferguson 1992, 224).

Conclusion

The underlying principle of Materialist theory is that warfare is uneconomical and dangerous; therefore it would only be employed if there were an acute need for food or territory (Thorpe 2003, 148). However, studies into materialist causes of non-state level warfare have revealed complex relationships between basic material needs and a variety of other factors, including the impact of colonial influences, population densities, socio-political organisation, cultural conditioning and the relative material expectations of those involved. While many agree to the multi-factorial nature of materialist interpretations of warfare, it is this last, the relative nature of material deprivation, which perhaps causes most problems. Ferguson's qualification that material lack should be measured on prior conditions and in comparison with similar communities (1984, 38) makes the scope of materialist interpretation unfeasibly broad. Hallpike voices concerns over this vagueness, suggesting that following such open-ended definitions it will always be possible to find someone for whom violence is advantageous 'ending with the extreme case of the functional value of murder for the killer, because it removes his inner tensions' (Hallpike 1973, 454). Another problem with Ferguson's hypothesis is his proposal that 'other motives [than materialist ones] offered by informants would be considered epiphenomenal, unless it could be shown that these other motives would, independent of material need, produce the actual pattern of fighting' (1984, 41). As mentioned above, it seems reasonable to assume economic reasons for many social interactions, warfare included, however, materialist until proven otherwise, is hardly a satisfactory argument. In reference to this proposition Robarchek asks "If I own a Volkswagen and I steal a Mercedes, is the cause of my behaviour 'material'?...the ontological assumption is made that only material causes are 'real'" (1990, 69). Ferguson proposes that the various interpretations intended to explain the motivations and causes of warfare are not mutually exclusive (1990, 54-55); for example he mentions that Vayda's cultural model, that repeated frustrations or deprivations result in aggressive responses, is both cultural (in that hostility is a socialised response) and materialist (as the reaction is an adaptive response to resource scarcity) (Ferguson 1984, 15). Again this is a reasonable conclusion, however, it is important not to underestimate the ability of culturally learned responses to reproduce themselves irrespective of material advantage, as discussed below.

2.4 Cultural Interpretations

Origins and Definition

Ferguson asserts that cultural approaches are the most widely cited of all interpretations of warfare in anthropology (1984, 14) and defines the cultural interpretation of warfare as the 'acting out of values and beliefs characteristic of a particular group' (Ferguson 2001, 99). The origins of this interpretation are difficult to establish beyond stating that before the emergence of materialist theories in the 1940s the main motivations for warfare in tribal societies were commonly thought to be prestige, sport or revenge (Ferguson 1984, 23; Maschner and Reedy-Maschner 1997). As a subset of multi-factorial approaches, this interpretation is usually seen as either a way of institutionalising values which can result

in material gain (Chagnon 1992, 93-95), for example, or as contributing to socio-political models of war as a means to develop cultural, social and political systems (Haas 1990; Carneiro 1990; Earle 1997). In anthropological studies, warfare as a mechanism for socio-political change usually takes the form of increased social complexity and a consolidation of dispersed groups. Such transformations are often cyclical in nature, as the 'progress' towards political complexity is often punctuated by periods of 'regression' in which groups periodically fragment into smaller polities. The processes involved have been noted by several anthropologists (Cohen 1984; Price 1984; Haas 1990; Earle 1997), having first been laid out formally by Robert Carneiro (1970) in his theory of circumscribed resources.

Carneiro's Theory of Circumscribed Resources

The underlying principle of Carneiro's theory of circumscribed resources states that warfare in circumscribed environments, in particular circumscribed agricultural land, such as islands, may result in dramatic rises in social complexity (1970). This occurs as the defeated groups cannot disperse and must be incorporated into the victor's community. Management of this increased population demands a more complex political organisation; and soon communities are locked into a cycle of war and increasing socio-political complexity, with each increase in population-density requiring more complex administration of people and resources. Carneiro illustrated the key manifestations of this theory with two case studies based on records of early Western contact with Fiji in the Southwest Pacific and the Cauca Valley, Colombia (Carneiro 1990, 193). In Fiji and Colombia, Western observers described numerous small, autonomous chiefdoms in competition over resources. Each of these chiefdoms consisted of a scattering of dispersed villages under the overarching control of one chieftain: a war leader who ruled by a combination of heredity and personal prowess. In both cases, this competition led to embedded patterns of warfare and an enculturation of warrior values, as political power was associated with military strength. Chieftains displayed actual or symbolic trophies of their defeated enemies around their palisades, and cannibalism and headhunting became a part of both secular and religious ritual practice (Carneiro 1990, 199-201).

War in these contexts was related to conquest, so that, in terms of long-term political growth, it resulted in taking control of a village or aggregation of villages, or, at a more advanced level, taking control of, and incorporating other chiefdoms, into states or proto-states. In the shorter term, however, the trajectory of this political evolution was marked by fluctuations. In Carneiro's two case-studies, success (the acquisition of power and resources) meant success in war, however, success in war had two important consequences. On the one hand, frequent warfare could result in increases in territory and people under one chiefdom. Contrarily, however, it also resulted in resentment which sometimes spilled over into the destruction, rather than the appropriation of enemy resources. Carneiro describes episodes of 'annihilation...[and] extermination' (1990, 200). These usually took the form of indiscriminate and needless destruction including demolishing every house in the village as well as the wasteful destruction of farms and food storehouses. These episodes often caused fluctuations, what Carneiro terms periods of 'progression' and 'regression' (Carneiro 1990, 207), as the progress towards the consolidation of several polities, was punctuated by episodes of regression with dispersed and fragmented settlement (Carneiro 1990, 207).

For Carneiro, the main factors actuating the ultimate consolidation of these chiefdoms into complex-chiefdoms or proto-states were the combination of population pressure and circumscribed agricultural land (Carneiro 1990, 192). The need for greater resources led chieftains to enter into wars of conquest, however, successful conquest also led to greater numbers of people. On islands or in areas of limited resources the defeated populations could not vacate their territory, necessitating more complex power systems to control the, now larger, polity (*ibid.*). Carneiro's model presents a convincing theory on some possible mechanisms driving increases in social complexity, however, while he describes how these polities are formed, he does not explain how and why they are sustained. Given that the population density and resources remain roughly the same before after the conflict, it may be just as likely that the chiefdoms would disperse into smaller, more sustainable polities rather than maintain the larger, more centralised structure.

Earle's Theory of Political Economy

Earle (1997) elaborates on the hypothesis that warfare can drive social change, by integrating the role of military force as a social evolutionary device, within the political economy as a whole. He proposes that in order to create lasting political centralisation, change must be driven by a synchronisation of military, economic and ideological control. According to this model, the initial manifestation of increasing complexity may be the cyclical rise and fall of chiefdoms (*ibid.*, 194). This early instability may signify power-bases which are emerging by force of arms, but which cannot sustain political control without consolidating their military power with economic and ideological authority. This appears to have been the case regarding the Wanka of Peru, who underwent almost a millennium (c. AD 500 – 1500) of conquest warfare which resulted in brief consolidations of power followed by the rapid collapse of the new polities and a resumption of small, fragmented chiefdom-level societies (Earle 1997, 194-97).

Likewise, control based on ideological and economic power can be difficult to sustain without military force. The Late Neolithic – Early Bronze Age societies of Thy, Denmark had a strongly defined elite, represented ideologically by items of personal wealth and by the burial monuments which reinforced the legitimacy of their lineages (Earle 1997, 129-330, 166). These chieftains made their wealth through the long distance trade of prestige goods and the trade-products derived from livestock. In both cases the economic basis for a chieftain's wealth would have been difficult to control by military means. A rival's trade could be disrupted, by, for example, cattle raiding, however, it would be difficult to co-opt such production through warfare over anything greater than a local area. International trade on the other hand was too wide-ranging to be controlled militarily. It was this lack of military capability which, Earle argues, kept power in Thy at a local rather than regional level (*ibid.*, 129, 197-200).

Earle presents the Kaua'i of the Hawaiian Islands (c. AD 800 – 1824) as a case study demonstrating the possible socio-political effects of the integration of military, ideological and economic power (Earle 1997, 111). According to their earliest oral histories, the Kaua'i appear to have engaged in frequent warfare, primarily as a way of expanding territory and resources. Kaua'i chiefs expanded their territory rapidly, initially at regional levels, but quickly appropriating other islands. Early in the course of these expansions, successful chiefs became associated with major deities, central to fertility, death and war (*ibid.*, 169-79). The principal source of power within these expanding chiefdoms was economic. Intensive cultivation led to a surplus of agricultural staples, creating a much more stable power base than, for example, the Thy chiefdoms, which relied on the vagaries of international trading systems. This agricultural economic power was the basis of Kaua'i dominance, creating necessary commodities to trade and a surplus to protect the chiefdom from fragmenting under economic or political stresses. However, the intensive agricultural strategies which fed this power were put into place rapidly and only after conquest warfare and ideological constructs had acquired and legitimised the power of the Kaua'i chieftains. By the time of European contact in the eighteenth century (Earle 1997, 36) the Hawaiian chiefdoms appear to have been approaching statehood-levels of political centrality (*ibid.*, 202).

For Earle, polities cannot increase in complexity unless they are based on stable power sources, and power sources cannot be stable unless they integrate military, economic and ideological authority. Warfare can acquire power and destroy competition, however, alone it is unstable and difficult to control. Ideology legitimises power, but is malleable, and can be co-opted and reinterpreted by new social orders challenging the established authority. Economy can provide a strong power base if it is based on desired products or staple goods, but in order to lead to political centralisation, economic control must be limited to one person or a small group of elites. Essentially, military, economic and ideological power must be synchronised in order to act as a force for sustainable, socio-political centralisation (Earle 1997, 203-208).

2.5 Conclusion

Most of the anthropologists cited here allow for multi-factorial approaches. There are many commonalities in each case (Ferguson 1984, 53; Ferguson 1990, 55; Haas 1990, xii) and differences can usually be seen as differences in degree or emphasis (McCauley 1990, 6-7). Biological interpretations explore the innate nature of violence both as a biological impulse and a potentially predetermined aspect of human social development. Nevertheless, questions regarding patterns of warfare or the socio-political outcomes of conflict relevant to this study are more usefully approached through materialist and cultural interpretations.

Material needs are usually considered in conjunction with other factors. The materialist subset 'cultural ecology' for example, examines ecological stress, population densities, and other factors of material scarcity, in tandem with 'cultural' influences, such as socio-political organisation or colonial impacts. Within this approach, the cultural aspects of a society (i.e. the specific history or political structures) may influence the pattern or form of warfare, but the underlying motivation in going to war is material. In the case of the Bisaasi-teri Yąnomami the breakdown in social structures intensified the conflict, however, the ultimate causes of warfare were materialist (the desire for new tools) and ecological (disease and ecological depletion).

Ferguson's (1992) case study is useful as it examines how far-reaching the impact of civilised-tribal contact can be. For archaeological interpretations of civilised-tribal zone conflict it may be apposite to mention briefly a study of Iban contact with nineteenth-century Western slave-traders (Gibson 1990). Slave-raiding was already a significant part of the Iban culture and economy prior to the arrival of European merchants. These external trade relationships, however, made drastic changes to internal patterns of warfare. In this instance, existing inter-tribal relationships of power and subjugation intensified as slaves ceased to be domestic forced labour and became economic commodities. The result of colonial impact among the Iban and their island neighbours was politically and socially significant. Levels of complexity within the various indigenous societies increased in relation to their degree of contact with the state-level markets. The Iban, however, were never 'Westernised' and the material manifestations of their growing power remained essentially indigenous (an exaggeration of traditional celebrations of warrior status and headhunting rituals).

Cultural theories interpret the causes and consequences of warfare in relation to each society's specific history and social structures. While materialists feel that a society's ideology, history or socio-political complexity may influence the form or extent of warfare; cultural theorists propose that these factors can instigate warfare and determine its outcome. In particular Earle's (1997) case studies demonstrate how warfare can be viewed as a social process rather than as a series of events outwith 'normal' social action. Although Carneiro (1970, 1990) and Earle (1997) propose different principal causes, their observations on the development of political complexity among chiefdoms, and the role that warfare plays in that

development, converge on many points. Most notably both describe fluctuations, periods of 'progression' and 'regression' in centralisation, which often accompany the development of complex polities. Carneiro and Earle also stress the importance of ideology in supporting military strength. Both of these things may, conceivably be identifiable in the archaeological record.

CHAPTER 3 THE ENVIRONMENTAL AND ARCHAEOLOGICAL BACKGROUND OF THE SOUTHERN FRENCH IRON AGE

3.1 Introduction

Lying between prehistoric, temperate Europe and the literate cultures of Greece and Italy, southern France was open to a variety of influences, from the Mediterranean world via coastal routes, and from temperate Europe through the river valleys of the Rhône and the Garonne. For most of the period between the late-seventh and early-first centuries BC, the rich and diverse archaeology of the region demonstrates aspects of all these influences but in many ways retains a distinct and independent character (Cunliffe 1988, 308; Garcia 2005, 168-77). The quantity and variation of Iron Age remains in southern France make a full examination of the evidence impractical within the confines of the present study. For this reason, the detailed analyses made in subsequent chapters will focus predominantly on the Region of Provence, and in particular on the Département of the Bouches du Rhône.

Many key Iron Age sites are found in the Bouches du Rhône (fig.1), as is much of the iconographic evidence associated with the *herôon*, or 'warrior sanctuaries' and the osteological evidence which has been interpreted as a 'cult of the head'. Of particular relevance to the study of conflict, this region is also within the immediate sphere of influence of Massalia, which played such a pivotal role in the advent of both Greek and Roman presence in the region. This permits closer examination of the direct effects on the indigenous communities of state-level, colonial intrusions of both the economic and militaristic sort.

3.2 Ecology

Topography

The southern coast of France runs for 500km along the Mediterranean Sea from Spain to Italy. Much of the hinterland is defined by an arc of mountain ranges which cut southern France off from temperate Europe. These mountains are breached by two great river valleys. The river Aude in the west creates a route to the Atlantic and western Europe. Further to the east, the Rhône, running from central France southward into the Bouches du Rhône, splits into two branches before flowing into the Mediterranean. West of the Rhône, the coastal regions of Languedoc-Roussillon are characterised by lagoons created by sand and gravel deposits from the Rhône, and large tracts of fertile soils (Cleere 2001, 3). The area east of the Rhône comprises three very different landscapes. The Camargue is a marshy region created by the delta of the diverging branches of the Rhône before it enters the Mediterranean. Like the coastal areas of Languedoc, it is full of salt and fresh-water lagoons, and the soil is enriched by alluvial deposits (Thompson 1975, 10-14). East of the Camargue is a bare steppe known as the Crau. Prior to modern agricultural improvements this region could only have sustained scrub for rough grazing (*ibid.*); an impression supported by the commentators Pliny and Strabo in the first century AD.

It is called the Stony Plain from the circumstance that it is full of stones as large as can be held in the hand, while beneath there is a wild grass which provides abundant feeding for cattle (Strabo IV, I, 7).

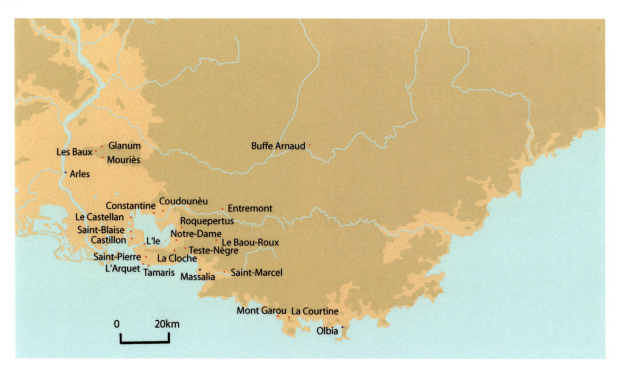

Fig 3.1 Main settlements mentioned in the text

Most of the archaeological remains examined in this study are concentrated in the Lower Rhône Valley around the Étang de Berre (fig 2). West and south of the Étang is the modern commune of Martigues. While most of the coastal and inland areas of this region comprise the same arid limestone of the rest of southeast Provence, the Plain of Saint-Julien (a depression or shallow valley rather than a plain) consists of 800 ha of fertile agricultural soil. To the north of Saint-Julien, in the so-called 'zone of étangs' the shores of the lagoons are enriched by the same alluvial deposits as the Camargue (Chausserie-Laprée 2005, 186). The landscape east of the Étang de Berre is an array of isolated, basins hemmed in by ranges of limestone hills (Thompson 1975, 10-11). Massalia was established in this area, and (although its influence was strongly felt throughout the Lower Rhône) within this study, the region east of the Étang de Berre will be referred to as the Massaliote hinterland. The uplands of this region were (and are) susceptible to erosion (Magnin 1993, Walsh and Mocci 2003). There appears to have been little woodland cover during the Iron Age and soils were regularly displaced by the dry climate and strong winds of the Mediterranean (below). These deposits built up in the low-lying basins creating fertile valleys which appear to have been in use by farmers since the Neolithic (Magnin 1993).

Climate

Pollen and coleopteran assemblages from Late Iron Age sediment layers near the Camargue, where most of the relevant palaeobotanical work in the region has been carried out, correspond strongly with current flora and fauna in the area. This suggests that the climate was similar to the present (Andrieu-Ponel et al. 2000a, 349-54), that is, typically dry all year round, with very hot and arid summers. Late summer and early autumn also see torrential storms in Mediterranean France. The region is sporadically beset by a violent northern wind of dry, freezing air from the Massif Central and the western Alps, known as the Mistral. This wind dehydrates the soil (Andrieu-Ponel et al. 2000a, 342) as well as being generally very destructive (Thompson 1975, 13). Called the 'Black North Wind' in ancient times, Diodorus Siculus, Strabo and Aristotle all record anecdotes of the Mistral hurling stones around the Crau and pulling men from their horses, stripping them of their clothes and weapons (Tierney 1960, 249, 258).

The climate, therefore, permitted the cultivation of a range of plants that could not have survived in temperate Europe; however, agriculture could also have been beleaguered by weather extremes. Most of the rainfall occurs during the spring and autumn (Dietler 1997, 274), at which times there are also often sudden frosts (coinciding with the harvest in the vineyards). Walsh and Mocci (2003, 55) also note that after clearance, the soils are subject to erosion as the top layers become dry and friable when unprotected by ground cover. In the summer storms, these ground surfaces can be washed downhill or simply blown away.

Food Production

Palaeobotanical analysis is rarely possible in Mediterranean France due to the poor survival of organic

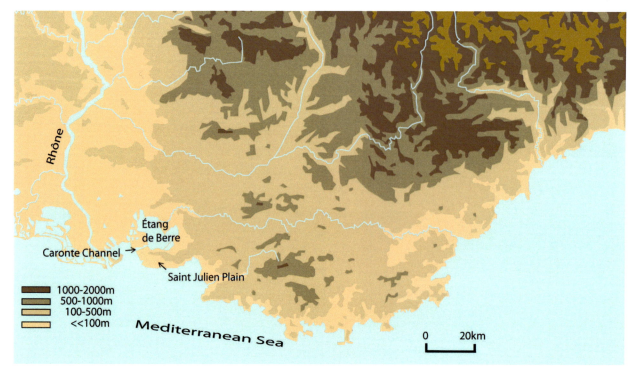

Fig 3.2 The Lower Rhône Valley: Martigues, west and south of the Étang de Berre and the immediate hinterland of Massalia to the east. Although generally arid with poor soils, pockets of alluvial deposits have built up in shallow valleys all over the Lower Rhône Valley. The largest of these is the Saint Julien Plain in Martigues. The shores of the Étang de Berre and the zone of étangs also provide relatively good agricultural soil. The Caronte Channel created a direct route from the Mediterranean to the heartland of early indigenous settlement.

remains, and any such researches generally focus on the transition to pastoral and agrarian farming in the Neolithic. The 'Iron Age' horizons of these studies usually reflect the Gallo-Roman period, which is noted due to the intensification of agricultural strategies after the annexation of the province (e.g. Trément 1999; Leveau 1999; Andrieu-Ponel et. al 2000a) The available palaeobotanical evidence for vegetation and agricultural practices suggests that southern France had seen much deforestation since the Neolithic, but that this increased during the Iron Age (Dietler 1997, 274). There is evidence for arable and pastoral farming from at least 750 BC (Andrieu-Ponel 2000b, 73), and in Languedoc the cultivation of olives, vines, walnuts and cereals are generally thought to have begun around this time (Andrieu-Ponel 2000a, 341; Buffat and Pellecuer 2001; Buxó 1996).

Most plant remains in the Bouches du Rhône have derived from settlements, preserved in burnt or waterlogged conditions in contexts of domestic cooking, processing and storage (Chausserie-Laprée 2005, 202). Wheat, the primary cereal in Languedoc, would have produced relatively poor returns in the less fertile soils of the Lower Rhône Valley. The waterlogged site of L'Île and the 'granary' farm of Coudounèu both suggest that in the Bouches du Rhône, the principal cereal may have been millet (*ibid.*). As in Languedoc vines appear to have been cultivated in Provence from the sixth century (Brun and Laubenheimer 2001). These may have been produced for eating or for grinding the pips for flour rather than for wine-making (Buxó i Capdevila 1996). From this period many of the accoutrements of supra-subsistence farming, such as granaries, silos and storage jars, seem to occur on settlement sites in the Midi (Garcia 2000, 55). By the fourth century (again at L'Île) crushed berries and must-residue suggest that grapes were grown for wine production (Chausserie-Laprée 2005, 204). The earliest micro-remains of olives occur in second-century BC levels, however, by the early-fourth century oil-presses appear frequently on indigenous settlement sites all over southern France (Chausserie-Laprée 2005, 204).

3.3 Iron Age Chronologies for France

The Iron Age communities of southern France have often been seen as belonging to essentially the same cultural milieu as societies further to the north. Modern national boundaries, with their foundations in the Roman conquest of Gaul in the first century BC, have compounded the impression of cultural homogeneity throughout Gaul. However, southern France, between the mountains and the Mediterranean Sea, differs substantially in climate and terrain from temperate Europe (Dietler 1997, 274) and it is perhaps unsurprising that its archaeology is also very different. The Hallstatt and La Tène chronologies which structure Iron Age chronologies for the rest of France are based on sequences and developments rarely evidenced in the development of material culture in the Midi. For this reason the Iron Age is often simply divided into early and late phases: the First and Second Iron Ages. These phases are subdivided into different

sequences to accommodate local or regional variations (*ibid.*, 275-77) (table 3.1).

3.4 History of Research

The Culture-historical Background
Much of the archaeological work done in the region, especially prior to the 1980s, followed a broadly culture-historical tradition in which three main linguistically-defined groups played a central role. Three languages were apparently spoken in the south of France at the time of the foundation of Massalia. Iberian was spoken west of the Rhône, Ligurian to the east and Celtic around the Lower Rhône Basin and the Aude (Dietler 1997, 274). Iberian, Ligurian and Celtic have also been used as ethnonyms associated with the region. Broadly speaking, the area east of the Rhône was seen as containing groups comprising Iberians, Ligurians and an amalgamation of both. A periplus (*Pseudo-Scylax* 3-4), referring to the populations of eastern Languedoc during the fourth century BC asserts that 'beyond the Iberians, Ligurians and Iberians live mixed up to the Rhône' (Dietler and Py 2003, 792). Around the Bouches du Rhône, the area with which this study is primarily concerned, Celts and Ligurians were referred to as distinct populations, while the term 'Celto-Ligurian' was apparently used by Strabo to denote an amalgamation of incoming Celts and the indigenous Ligurians (Benoît 1975, 227).

One of the earliest texts describing these communities comes from Hecataeus of Miletus, writing around 500 BC. Hecataeus described the cities of southern Gaul by reference to the indigenous inhabitants rather than the colonists who established the settlements (Pearson 1939, 38). Hecataeus recorded that 'Narbo [Narbonne] is a Celtic city and Massalia is a Ligurian city to the east and south of Celtic territory' (54-55). Almost 600 years later, in the first century AD, Strabo continued to make this distinction between the two groups: 'Many tribes occupy the Alps and all are Celtic except the Ligurians; they belong to a different race, but their way of life is similar to that of the Celts' (II, 5, 28). However, the precise meaning of these differences, 'racial' or otherwise, is not explained.

Ligurian is a vague appellation which has elicited various explanations from the textual sources. As mentioned above, Hecataeus associated Ligurians with southern France from at least the foundation of Massalia (54-55). An early reference appears to have survived in a late text. Festus Rufus Avienus, writing his periplus *Ora Maritima* in the fourth century AD drew on material from the Himilco periplus, which described the coastal areas of southern Gaul. The accounts of the harbours along the Mediterranean Sea in this document were minutely documented, however, the entrepôts of Emporion and Rhode, dating to the early sixth century BC, were not mentioned. This suggested that the early periplus pre-dated the foundation of these trading posts, and originated in the very early sixth century (Tierney 1960, 193). Avienus reported that Ligurians lived in northern Spain and the Iberian Peninsula, and mentions that they abandoned regions in northern Gaul which had been flooded. This account suggested to the French historian and Celtic scholar, Henri d'Arbois de Jubainville, working in the late nineteenth and early twentieth centuries, that Ligurians were a dominant presence all through western Europe prior to the 'arrival' of the Celts (Collis 2003, 63-66).

The term 'Celtic' is particularly problematic. The first references to Celts or 'Keltoi' come from Greek sources which referred to populations in various areas of France and Spain (Armit 1997, 13). As mentioned above, Hecataeus writing in the late sixth century BC mentioned that Massalia was established in lands occupied by Ligurians and Celts. Celts are also referred to by Avienus who, as mentioned above, may have drawn on material from the early sixth century (Tierney 1960, 193). Avienus also referred to Celtic tribes at this time throughout France and Spain (*ibid*). In the fifth century BC, Herodotus also placed the Celts not only near the Pyrénées, but further north, in the Iberian Peninsula and central Europe (Tierney 1960, 194; Cunliffe 1997, 3). Collis proposes that Herodotus's statement that the Danube has its origins in the land of the Celts near a town called 'Pyrene' had been based on a misapprehension of western European geography. He suggests that 'Pyrene' was in fact the town of 'Portus Pyrenaei', recorded by Livy as being to the north of Rhode. Collis believes that Herodotus was recording the presence of Celts in southern France around the Pyrénées (Collis 2003, 126). Herodotus, by his own admission, appears to have been unfamiliar with the western regions of Europe (Pearson 1939, 37).

One of the few Mediterranean writers to have had first hand knowledge of the Celts was Poseidonius, who travelled through southern Gaul in the late second century BC. Although his books did not survive, the writings of Poseidonius formed the basis for works by Strabo, Diodorus Siculus and Caesar in the final century BC (Tierney 1960; Piggott 1968, 83-84). Strabo and Diodorus both associated the Celts with southern and central Gaul (Tierney 1960). Strabo stated that 'the Celtae [are] those who inhabit the southern parts of the Pyrénées, down to the sea off Marseilles and Narbonnes' (IV, I, 1). Diodorus Siculus drew a distinction between 'Celtae' (Celts), a specific name for the populations of southern Gaul, and 'Galatae' (Gauls), a generic term for the tribes occupying the rest of temperate Europe as far east as Scythia:

Those who live in the interior above Marseilles, and those along the Alps and those on this side of the Pyrénées are called Celts, whereas those who are settled above Celtica in the area stretching towards the north…and all the people beyond them as far as Scythia, are called Galatae; the Romans, however, include all these tribes together under one name and call them all Gauls (V, 25, 32).

Diodorus's account appears to be corroborated by Caesar who writes that a third of Gaul is inhabited by people:

"called in their own tongue Celtae, in the Latin Galli" (De Bello Gallicum, 1:1). Diodorus also reflects the contradictions in the Greek and Roman texts, some of which locate the Celts directly around the Alps and the Pyrénées, while others extend Celtic geography north and west into central Europe and the Iberian Peninsula. It seems clear that the people associated with the name Celtae or Galli by Mediterranean authors, can at least be assigned to the area of southern and central France around the Pyrénées, even if their location in other parts of Europe is equivocal. The term 'Celtic', however, has other associations besides a name in the histories and ethnographies left by the Greeks and Romans.

Hallstatt and La Tène artefacts from west and central Europe have conventionally been interpreted as the material remains of the Celts so described in Mediterranean texts. Celtic in its modern sense, however, is usually understood to be a linguistic term, referring to a family of Indo-European languages probably originating in the Bronze Age. Celtic languages and La Tène artefacts are widespread throughout Europe, and while they often occur in the same areas, the two are not coterminous (Sim-Williams 1998; James, 1999).

Although Celtic was spoken in the region during the Iron Age (Dietler 1997, 274), there are relatively few examples of La Tène material culture in southern France, however, those traces found have conventionally been interpreted as the recognizable influx of material culture equating to the 'arrival' of Celts. Modern approaches highlight the need to be cautious in assigning ethnic names to archaeologically-defined groups, such as La Tène, in the region. There is no indication that the people covering the wide swathe of Europe referred to as 'Celtic' in the written sources thought of themselves in terms of cultural or ethnic unity or even saw themselves as 'Celts'. The repeated allusions by classical authors to Celts in the vicinity of Massalia and southern France, bolstered by Caesar's statement that they call themselves 'Celtae', may suggest that the populations of southern France may have considered themselves Celts, however, references to the Celts or Celto-Ligurians of southern France, do not carry any implied cultural or ethnic links with the La Tène using cultures further to the north

Early Investigations: The Art-historical background

Many of the major Iron Age sites and artefacts of southern France were discovered during the nineteenth and early twentieth centuries. During this period the influence of classical studies led to an essentially art-historical approach to archaeological investigation. Interpretations were largely based on possible architectural influences from the state-level societies of the Mediterranean. Southern France in particular was susceptible to this approach as the region had been the focus of Greek and Italian presences much sooner than the rest of Gaul. Caesar, for example, writing in the first century BC described the courageous and warlike nature of other Gauls in contrast to 'the culture and the civilization of the Provence' (De Bello Gallicum, 1, 1).

The pitfalls of this art-historical approach are exemplified in the interpretation of three blocks of stone sculpture depicting human heads and horse riders uncovered on the oppidum of Entremont in 1817. After exploratory excavation the publication of the sculpture and the site in Statistique des Bouches du Rhône proposed that the site was the Roman encampment of Marius. Subsequently, however, Entremont was recognized as the principal political centre of the Saluvii, an indigenous Iron Age population in Provence (André and Charrière 1996). This classical studies approach also continued an earlier antiquarian tradition, which emphasized the retrieval of specimens for collections. For instance the two initial excavations at the sanctuary at Roquepertuse in 1919-24 and 1927 by Gérin-Ricard concentrated on the recovery of the monumental statuary and a proposed reconstruction of a three-pillared portico. Interpretations of the possible economic, social or ritual aspects of the site were not attempted (Lescure and Gantès 1991, 17).

Access to documentary evidence encouraged interpretations that relied heavily on historical accounts, often resulting in the equation of material culture with the historical events or ethnographies recounted by Greek and Roman writers. For example, the periplus of Himlico, which dates to around 500 BC, makes reference to Celts only along the Atlantic coast (Collis 2003, 66). However, Polybius's account of Hannibal crossing the Pyrénées in the third century BC, mentioned the presence of Celts in southern France (*ibid*, 124). These two documents formed the basis for the interpretation developed by the late nineteenth and early twentieth centuries by d'Arbois de Jubainville, which proposed that incoming Gauls subjugated the indigenous, southern French populations around 300 BC (Collis 2003, 66).

It is likely that Fernand Benoît, who conducted excavations at Entremont from 1946 until his death in 1969 (André and Charrière 1996), was building on this model in his interpretation of the site. Benoît identified the native population of southern France at the time of the initial Greek settlement as the Ligurians, and attributed the earliest phase at Entremont to the period prior to the 'arrival' of the Celts (Benoît 1975, 245). These dominant incomers merged with the native Ligurian population, and the resultant 'Celto-Ligurians' formed a powerful military alliance, the Saluvii, uniting all the tribes from the Rhône to Maures mountains. In this scenario, finds of 'Celtic' metalwork such as La Tène fibulae and belt buckles found at Entremont (*ibid.*, 239), became bound to the historical narrative and were taken as proofs of population movements. More recently, authors have suggested that care should be taken with such terms as 'Celtic' or 'Ligurian' that might have had any number of meanings (ethnic, political, geographical etc.) (e.g. Hodson and Rowley 1974, 191).

Perhaps the dominant theme in mid-twentieth century studies of the southern French Iron Age, however, was that of Hellenisation. Benoît's interpretation of the oppidum of Entremont, for example, centred on the degree to which contact with Mediterranean colonists influenced native traditions in architecture and religion. Coming from a background of classical and historical studies, the archaeologists of the early twentieth century were inclined to view the changes in native settlement patterns as reflecting a Hellenistic, 'civilizing' influence on native populations. More recent approaches prefer to see the impetus for social change as coming from within indigenous historical and political organisation (Dietler 1995, 65).

Together these three perspectives, art-history, culture history and the Hellenisation model, encouraged the use of archaeology as corroborative evidence for early written accounts, equating material culture with people, and interpreting changes in native archaeology as essentially the imposition of Mediterranean cultural values or the result of Celtic migration. More recently, in part influenced by New or processual archaeology, archaeologists in the region have begun to move away from reliance on historical explanations and instead have attempted to reconstruct how social and economic systems might operate and what causes them to change.

Modern Excavation and Research

In recent years the archaeology of southern France has benefited from the development of these new interpretative approaches. During the Iron Age, increasingly complex economic and social systems are evidenced in the growing magnitude and sophistication of ritual structures and the size and complexity of settlements and administrative centres in the region. Recent studies view these developments, not only as a reaction to contacts with the Mediterranean world and temperate Europe, but also within an indigenous context. There is no doubt that the Mediterranean world played a key role in several of the transformations seen in southern France during this period, but the nature and extent of this interaction is complex, and a simple verdict of 'Hellenisation' is not appropriate (Kristiansen 1998, 327-29). Modern approaches recognise the need for caution when making comparisons with the rest of the Celtic world as well as with the other Mediterranean cultures (Arcelin et. al 1992, 181).

The other significant development of recent years is the increasing use of scientific analyses, for example, the examination of methods of stone cutting in the construction of monumental architecture in Iron Age Gaul (e.g. Bessac 1991, 43-53) and the sourcing of pigments in the painted statuary (e.g. Barbet 1991, 53-81). Scientific approaches have also highlighted the limitations of previous work, such as the reconstruction of structures at Roquepertuse, mentioned above. Recent reappraisals of the material from this site have prompted suggestions that very little attention was paid to ascertaining the exact original locations of the reconstructed fragments, and even to the technical credibility of the reconstruction (Lescure and Gantès 1991).

3.5 The Iron Age in Southern France: A Brief Summary

Early Indigenous Settlement

There was some continuity between the end of the Bronze Age and settlement in the First Iron Age (c. 750 - 700 BC to c. 475 - 450 BC) in southern France. In the mountainous regions of eastern Provence, rock shelters and caves continued to be used, as did small hilltop settlements. In the lowland plains and coastal regions of the Lower Rhône and Languedoc small villages clustered around the edges of lagoons and along the shoreline (Dietler 1997, 310). Houses, usually one-roomed sub-rectangular structures, were built with local materials. Those in the upland, mountainous areas used stone, while the majority of settlements, those on the shores and alluvial plains, were timber framed with wattle and daub or cob walls (Cleere 2000, 19; Garcia 2000, 51). Defensive features, such as ramparts, remained very rare both at the end of the Late Bronze Age and throughout the Early Iron Age (Dietler 1997, 310-13; Audouze and Büchsenschütz 1989, 81).

In the late eighth century the advent of iron tools in southern France introduced the potential for more intensive land-clearance and agricultural production. The earliest evidence for Mediterranean trade occurs around the same time (Garcia 2000, 51). By the late seventh century, the earliest permanent, enclosed indigenous settlements were established. The early centres at Saint-Blaise in Provence and Lattes in Languedoc have yielded large quantities of Etruscan ceramics suggesting they were centres of trade. Dietler (1997, 277-79) states that no evidence of permanent Etruscan settlement has yet been found, although recent excavations at Lattes may prove otherwise (Thierry Janin pers. comm.). In around 600 BC the permanent Phocaean-Greek settlement of Massalia was founded on a natural harbour in Provence, overlooking the Bay of Lions.

The Emergence of Indigenous Urban Settlement

The sixth and fifth centuries saw a dramatic change in indigenous settlement patterns. These centuries, known as the 'proto-urban' period (Chausserie-Laprée 2005, 44) witnessed the establishment of numerous enclosed indigenous agglomerations along the coast, and particularly around the Lower Rhône. The surrounding walls or ramparts and a degree of overarching conception in internal planning imply a level of centralised political control. A striking concentration of such centres is in evidence around the region of Martigues (Chausserie-Laprée 2005, 42; Dietler 1997, 313). Here, the great variations in settlement size suggest a certain level of hierarchy. The duration of these centres was also variable. Some were occupied for centuries while others were abandoned after a short occupancy (Chausserie-Laprée 2005, 42). Further inland meanwhile indigenous settlement appears to have continued much as it had done in the Late Bronze Age (Dietler 1997, 113; Walsh and Mocci 2003, 47).

Greek Expansion and the Consolidation of Indigenous Urbanisation

Historical sources state that subsequent to the foundation of Massalia, secondary Phocaean colonies were established throughout the fifth and fourth centuries at Agde, and Olbia. The presence of Greek ceramics at indigenous sites all along the Gulf of Lions, such as Mailhac, Pech Maho and Arles, have prompted the suggestion that enclaves of Greek traders had involved these settlements in Massaliote trading networks, however, the exact mechanism of this relationship is far from clear (Dietler 1997, 287). Nash also mentions that, given the Phocaean reputation for piracy, there is no reason why the colony would not have visited the coastal settlements of France to obtain plunder and slaves for the Mediterranean market (Nash 1985, 56; Goudineau 1983, 81). It is possible that, even from its earliest phases, colonial activity could have had a violent and destabilising impact on the region.

The Second Iron Age signalled a transformation in indigenous settlement patterns. Throughout the fourth and third centuries, the indigenous, lowland settlements were abandoned in favour of small, defensive, upland villages (Arcelin et. al 1992, 181-82). These centres were concentrated around the immediate hinterland of Massalia. The settlements were generally smaller (seldom more than 1 ha, but they adhered to a much more precise layout (Chausserie-Laprée 2005, 46) and are often viewed as evidence of the consolidation of the indigenous urban tradition (Garcia 2004, 80-89).

The Emergence of Complex Oppida and the Advent of Roman Rule

In the late third and early second centuries, many of these settlements were destroyed and abandoned. After a further brief episode of dispersed settlement, the population subsequently became much more concentrated, and many of these sites, such as Entremont in Provence, underwent a period of reoccupation. In these oppida, or hillforts, domestic units, ritual spaces and industrial or craft-working zones, were formally laid out and enclosed by very substantial bastions (Arcelin et. al 1992, 181-82; Dietler 1997, 314-16). The degree of Greek influence in these indigenous developments is the subject of much debate. Some see these constructions as Greek technology adapted to native ritual expressions, such as the numerous structures devoted to the 'head cult' (Arcelin et. al 1992, 183; Goudineau 1983, 84). Others describe Entremont, with its monumental architecture, religious sculptures and streets laid out in grids, as being "hardly distinguishable" from Greek Massalia (e.g. Collis 1995, 163).

From the third century BC a degree of Roman influence in southern France can be seen in the increasing importation of ceramics associated with wine drinking. This trend appears to have increased throughout the second century, as evidenced by concentrations of Dressel 1a wine amphorae and black glazed wares all along the Rhône Valley (Fulford 1985, 94). Unlike Etruria and Greece, however, these economic incursions would eventually develop an imperialistic objective. With the establishment of the Roman enclave of Narbo Martius (Narbonne), Rome began a campaign of 'pacification' and subsequently, in the late second/early first century BC, established political and administrative control in the region (Dietler 1997, 291-93).

It is clear that throughout this period, southern France underwent a period of considerable social and cultural transformation. The first permanent Greek colony, Massalia, was established by Phocaeans around 600 BC, a date supported by recent excavations (Dietler 1997). After this time, indigenous settlement patterns seem to have fluctuated between episodes of concentration and dispersal. Each period of expansion, however, displayed an increase in scale and uniformity, suggesting a growing trend towards centralisation. Ritual expression, especially in terms of stone sculpture also underwent changes, which adhered to indigenous themes, while developing an increasingly monumental and urban form. These developments were clearly related to the changing and complex relationships with the Mediterranean world, which impacted upon the indigenous population of southern France in a variety of roles: as neighbours, trade partners, colonizers and, eventually, conquerors.

3.6 Conclusion

The period between the late seventh and the first century BC was a time of upheaval in the south of France. Models which see warfare as a mechanism for the development of social and political complexity (such as the broadly processual models of Carneiro 1978, 1990; and Price 1978, 1984), suggest that these changes may have come from within the indigenous sphere of control. However, conflict, both with outsiders and between different sections of an indigenous population, is often precipitated by contact between states and non state-level societies (Gibson 1990). Italian and Greek states became involved with southern French society through both mercantile and military colonialism. Social anthropologists have discussed how 'asymmetric culture contact' in any form, can greatly destabilize indigenous systems of organisation (Price 1978, 166). The Iron Age of southern France provides the opportunity to consider many issues related to the effects of contact between societies of different social-political complexity, including increases in centralisation, destabilising processes (e.g. slave-raiding) and encultured warlike behaviour. The documentary evidence examined in the following chapter will reflect different aspects of the archaeological changes in southern France outlined above, from the point of view of these colonial powers.

COLONIAL EVENT	BC	Louis et. al 1968 Western Languedoc	Py 1993 Mediterranean France	Arcelin 1976 Province
	100	Late Iron Age	Fer II Final	Phase V
Caesar's conquest of Gaul				
Roman conquest of Med. France	200		Fer II Recent	Phase IV
	300		Fer II Ancien	
Olbia founded				
Agde founded	400		Transition Fer I/II	
	500	Period IV Grand Bassin II		Phase III
Emporion founded			Fer I Recent	Phase II
Massalia founded	600	Period III Grand Bassin I	Fer I Ancien	Phase I
	700		Transition Bronze/Fer	
Etruscan trade		Period II Moulin		Bronze Final
	800		Bronze Final IIIb	

Table 3.1 Iron Age Period Systems for Mediterranean France: a selection of different chronologies showing Mediterranean France as a whole, and two proposed sequence of southern French archaeology west and east of the Rhone. In southern French archaeology the divisions 'First' and 'Second' Iron Age are commonly used. The First Iron Age is the period from the late eighth century until the early/middle fifth century. The Second Iron Age is the period from the mid fifth century until the Roman annexation.

CHAPTER 4 THE DOCUMENTARY EVIDENCE

4.1 Introduction

The first strand of evidence to be considered in this study is the collection of written records left in the wake of the early contact between the native peoples of Iron Age southern France, and the literate societies of Greece and Rome. This chapter begins by compiling a 'chronicle' of the Iron Age in the region, particularly focusing on incidences of conflict, and attempting to gauge the reliability of the sources and any archaeological evidence for the events described. Part two contemplates several of the more relevant ethnographic accounts from southern France. It is possible that many of the characteristics ascribed to the stereotypical 'Celtic warrior' derive from Provence, and some of the practices described, such as headhunting, are briefly examined to ascertain how their physical vestiges may suggest warfare and/or ritualised violence in the indigenous societies of that region, as well as reflecting the often strained relationship of the Gaulish-Mediterranean cultural tribal-zone.

4.2 Part I: The Historic Accounts

The Sources

The earliest written sources concerning southern France were of Greek origin, following colonisation along the Mediterranean, and the establishment of Massalia c. 600 BC (table 4.1). In their accounts of this first contact, the Greeks used historical, ethnographic and geographic narratives to illustrate their world-view. Accordingly, whether sympathetic or censorious, each account was coloured by the writer's particular philosophy (Rankin 1995, 27). In the fourth and third centuries, tribes moving south across the Alps carried out sporadic assaults on Greek and Italian communities in a series of events commonly referred to as the 'Celtic migrations'. Around the same time, documentary evidence indicates friction between the colony at Massalia and the indigenous Gaulish population. The ensuing volatility appears to have continued for the next two centuries (Justinus, *Historiae Philippicae* 43.5; Silius, *Punica* XV, 169-72). Reports of foreigners in general and Celts in particular became more critical, focusing on their more bizarre or distasteful customs. Eventually, however, this 'hard primitivism' softened, as those living beyond the civitas evoked a Golden Age of innate, natural simplicity in contrast to the luxurious and corrupt living enjoyed in the city-states of the Greeks and Romans (cf. Lovejoy and Boas 1935).

Latin sources began later, and the approach of Roman authors differed somewhat to the Greek tradition. Roman accounts of the Gauls did not follow the custom of philosophic enquiry employed in the Greek ethnographies (Rankin 1995, 27). Rome was an emerging state power at the time of the 'Celtic' migrations. Initially these assaults resulted in much damage and loss, with Rome itself overrun in 391-90 (Justinus 43.3). In the third century BC, however, the Battle of Telamon signalled a decisive coup against the 'Celtic' threat. Despite this, it has been postulated that Rome never fully relinquished its anxiety regarding this former enemy (Rankin 1995, 24). It is perhaps little wonder then that the Roman sources, whether historical or ethnographical in genre, are largely militaristic and propagandist in nature.

A brief caveat regarding chronology in all the sources quoted in this chapter is also necessary at this point. As mentioned above, chronologies based on texts dating to before the sack of Rome at the end of the fourth century BC were lost in the attack. The surviving documents were correlated by P. Mucius Scaevola in 133 BC (*Annales Maximi*). Collis notes that Cicero used this source which describes an eclipse of the moon in 351 BC but which is known to have taken place in 354 BC, thereby suggesting inaccuracies in the earlier dates (*ibid.*, 18 - 19). Accordingly, the annals record the sack of Rome as occurring in 390 BC, whereas Polybius, who takes his date from a historical reference (the battle of Aegospotami) dated the sack of Rome to 387 BC (*ibid.*, 19). Inaccuracies in the early dates, and discrepancies where some authors have used the Roman annals and others taken their chronologies from certain historic events, means that inevitably the dates and sometimes even the order of events will be fraught with inconsistencies. Moreover, there is every chance that, because multiple sources were used by each writer, authors might even have been using different chronologies within one account (Walbank 1972, 106). The inaccuracies of chronology are too convoluted to attempt redress here, but are simply recounted as another limitation of the textual evidence.

Sixth Century BC: The Foundation of Massalia

The earliest reference to southern France and its populations derives from the Himlico periplus which partially survives in the fourth century AD works of Avienus. The periplus, which may date to the very early sixth century BC, mentions Celts and Ligurians along the southern French coast (Tierney 1960, 193). More specific reference to the Celtic and Greek inhabitants of southern France comes from the Greek geographer Hecataeus (c. 540 – 475), however, once again his work survives only in fragments and quoted by a much later author, Stephanus of Byzantium (c. AD 480 – 479) (Koch 2000, 5; Collis 2003, 16). According to Stephanus, Hecataeus wrote that Narbon was a trading centre and city of the Celts (Europa Fr. 54), and that Massalia was a Phocaean colony in the land of the Ligurians, near Celtica (Fr. 56). Written records, therefore, appear to begin with the founding of the Greek colony of Massalia and, naturally, the authors were very often concerned with the relationship between the colonists and the native inhabitants.

Ancient sources proffer two dates for the foundation of Massalia. Pompeius Trogus (late-first century BC) and Livy (64/59 BC – AD 17) place the foundation of Massalia in the reign of Tarquin the elder (658 – 578 BC). However, Timaeus (352 – 256 BC) held that the

CENTURY	SOURCE	SOUTHERN FRANCE	ELSEWHERE
6TH CENTURY BC **600 BC**	HECATAEUS OF MILETUS (Fr. 53-56)	GREEK COLONISATION FOUNDATION OF GREEK-PHOCEAN COLONY OF MASSALIA	GREEK AND ETRUSCAN COLONISATION MASSALIOT TRADE PENETRATING 'HALLSTATT' EUROPE
5TH CENTURY BC **c450 BC**	POMPEIUS THUCYDIDES SOSYLOS JUSTINUS STRABO	VAGUE REFERENCES TO CONFLICTS INVOLVING MASSALIA	ROME AND ETRURIA VIE FOR SUPREMACY IN ITALY END OF MASSALIOT TRADE WITH 'HALLSTATT' EUROPE
4TH CENTURY BC **391-390 BC** **390 BC**	JUSTINUS (43.5) (*Hiestoria Philippicae*) JUSTINUS (43.3)	SIEGE OF MASSALIA: Greeks overcome Gallic attack	'CELTIC MIGRATIONS' Population movements across Alps into north Italy, southern France, Spain and Asia Minor SACK OF ROME
3RD CENTURY BC **225 BC** **219 - 202 BC**	SILIUS (XV, 169-72) (*Punica*) POLYBIUS (2.28. 3-10) (*Historia*) Q. FABIUS PICTOR POLYBIUS LIVY	GENERAL UNREST: Between the citizens of Massalia and native populations SECOND PUNIC WAR: Hannibal marches through the region enlisting mercenaries	 BATTLE OF TELAMON: Rome defeats Celts in north Italy SECOND PUNIC WAR: War with Carthaginian Spain
2ND CENTURY BC **154 BC** **125 - 124 BC** **?123 BC** **106 BC** **102 BC**	STRABO (33.7-11), LIVY (*Epitome*, LX) STRABO (IV, I 5) (*Geographica*) STRABO (4.1.13), JUSTINUS (3.3.36) STRABO (IV, I 8)	QUINTUS OPIMIUS' CAMPAIGN: the first of a series of campaigns by the Romans against Gauls and Ligurians... M. FULIUS FLACCUS' CAMPAIGN SEXTIUS FOUNDS CITY (AIX) NEAR MASSALIA CAEPIO'S CAMPAIGN FOUNDATION OF THE PROVINCE MARIUS'S CAMPAIGN	PERIOD OF ROMAN EXPANSION: provinces in Spain (195 BC) and Africa (146 BC) added to Sicily and Sardinia THE THIRD PUNIC WAR (149 - 146 BC) Strabo records that military campaigns were carried out by Rome in southern France to keep the route to Spain free.
1ST CENTURY BC **58-52 BC** **49 BC**	 CAESAR (*De Bello Gallico*) STRABO (IV, I, 5)	GALLIC WAR: The Helvetii attempt to pass through southern France ROMAN CIVIL WAR: Massalia supports the losing side, thus losing its autonomy	 GALLIC WAR ROMAN CIVIL WAR

Table 4.1 Framework of events

settlement was established 120 years prior to the battle of Salamis, that is to say 600 BC (Rankin 1996, 35). While there is no explicit account of any negotiations involved in instigating the colony, there are late versions of a foundation myth, presumably rooted in the need to legitimise the foundation and/or emphasise a relationship of welcome and integration with the indigenous population.

The account given by Pompeius Trogus survived in the writings of Justinus (second / third century AD) (Collis 2003, 23-4). In this version the young Phocaeans, Simos and Protis voyaged to Gaul where they met Nannus, the king of the Segobrigii and petitioned for an alliance. Their arrival coincided with the wedding of the king's daughter Gyptis, who, in accordance with tradition, would select a husband at a feast held the same day, by offering water to one of those present. Rejecting all her native suitors, Gyptis brought the water to Protis. 'He who had been a guest was now a son-in-law; and he obtained from his father-in-law the land on which to found a city' (*Historiae Philippicae* 43.3). In *The Constitution of Massalia* Aristotle (384 – 322 BC) gives a similar account which was preserved in the writings of Athenaeus (*Deipnosophistae* 13.576) some time after AD 193 (Koch 2000, 39; Collis 2003, 24). Once again the extant version dates to around eight hundred years after the supposed event.

Fifth century BC: Gauls, Ligurians and Carthaginians

In contrast to the sentiments of amity embodied in the Protis myth, Pompeius went on to make reference to early hostilities involving Massalia. Justinus omits the details of these conflicts, saying only that, according to Pompeius, Massalia had defeated Gauls and Ligurians and won a victory over the Carthaginians. He also mentions that, in all her wars, Rome had been supported by Massalia (43. 5), a relationship which would greatly influence the history of southern France throughout this period. The Gaulish and Ligurian conflicts are obscure seem related to the initial establishment of Phocaean territory in the region, as according to Justinus (43.3.13):

Having defeated their enemies, they established many colonies in the conquered lands

Strabo records with a little more detail (4.1.5):

They followed their natural inclination for the sea, but, later, their valour enabled them to take in some of the surrounding plains, thanks to the same military strength by which they founded their cities

Ebel proposes that the Carthaginian defeat might refer to a rather more specific event, and that a possible date for the battle might even be suggested by reference to other sources. Thucydides (1. 13) alludes to a Massaliote victory over the Carthaginians in the late-sixth century, while Sosylos appears to mention a similar event in the very early fifth century (490 BC). Ebel argues that coming after the Carthaginian defeat at Himera, Sicily in

480, an alliance between the Massaliote colonists and the Iberians was quite likely. He also points out that Pompeius's history appears to be written chronologically, and that, subsequent to the above events, he describes the sack of Rome in 390 BC (1976, 10-11). In any event, it would appear that between the late-sixth and early-fourth centuries, Massalia was engaged in, at least periodic, conflict with her immediate neighbours, and with populations further to the west, either in defence of her own daughter colonies, or in response to appeals from Rome.

Fourth century BC: The 'Celtic Migrations'

The arrival of the Greeks might have been expected to precipitate a fully-formed historical record concerning the fortunes of the colony and its relationship with the indigenous population; however, there is no extant *Constitution of Massalia*. The earliest instances of specific historical events concern hostilities involving the Greeks and their Ligurian and Gaulish neighbours around 200 years after the foundation of Massalia. These conflicts took place against a backdrop of increasing instability as Rome vied with Etruria as the emerging power in Italy, and populations from central and/or western Europe spread southwards into northern Italy and southern France. Reference is made from this time to a relationship of mutual support between Rome and Massalia. Several sources offer accounts of the taking of the Etruscan city of Veii by Rome in 396 BC, and although an alliance is not overtly mentioned, after the subjugation of it's long-term rival, Rome dedicated a gold krater at the treasury of Massalia at Delphi (Diodorus 14.93.3; Livy 5.28.3).

The Sack of Rome 391-90 BC

Throughout the fifth and fourth centuries BC, population movements appear to have taken place from 'Celtic' areas of temperate Europe, across the Alps into Spain, southern France, northern Italy and Asia Minor. This exodus was undoubtedly one of the defining events of the period, however, the evidence is elusive. As discussed below, there appears to be some archaeological confirmation of these migrations, however, extrapolation of ethnicity from material culture is clearly problematic. Collis states that, as regards the Celtic migrations, 'independent historical corroboration' is necessary to clarify the relationship between the archaeological remains and their ethnic associations (1984, 138). Unfortunately, literary references to the migrations are vague, the sack of Rome in 390 BC being one of the few explicit events recorded (Ebel 1976, 64-65). Pompeius' version of the assault on Rome is, once again, much abridged by Justinus. He recounts how the Gauls escaped overpopulation and internecine conflicts by sending 300,000 men to find new territory:

They expelled the Etruscans from their lands and founded Milan, Como, Brescia, Verona, Bergamo, Trento and Vicenza (XX 5.7-9).

Some of these settled in Italy, where they captured and burned even the city of Rome; some, led by the birds...spread through the head of the Adriatic and settled in Pannonia (XX 4.1-3).

Pliny the Elder (AD 23-79) is also brief, but offers an alternative motive:

...a Gallic citizen from Switzerland named Helico, who had sojourned at Rome on account of his skill as an artificer, had brought with him when he came back some dried figs and grapes and some samples of oil and wine; and consequently we may pardon them for having sought to obtain these things even by means of war (*Naturalis Historia* XXII 2. 5).

The most comprehensive and least simplistic version of the events surrounding the sack of Rome is supplied by Livy (*Histories*, V. 33. 2-6). He offers a combination of causes for a large-scale migration, including over-population and the lure of Mediterranean 'luxuries', and adds a dash of political intrigue:

Allured by the delicious fruits and especially the wine - then a novelty - [the Gauls] had crossed the Alps and possessed themselves of lands that had once been tilled by the Etruscans; and that wine had been imported into Gaul expressly to entice them, by Arruns of Clusium, in his anger at the seduction of his wife by Lucumo. This youth, whose guardian he had been, was so powerful that he could not have chastised him without calling in a foreign force...[but] those who besieged Clusium were not the first to have passed the Alps...indeed it was two hundred years before the attack on Clusium, and the capture of Rome, that the Gauls first crossed over into Italy.

According to Livy, the emissaries sent from Rome to negotiate the cessation of hostilities at Clusium behaved badly and, far from bargaining a settlement, one of the 'diplomats' killed a Gaulish leader. The party were routed and the Gauls decided to march southward to Rome. Livy (5.41) went on to convey the tragic dignity with which the Romans met their fate, and the devastation which followed:

...the nobles were massacred where they sat. No one was spared as the houses were ransacked and burned to the ground

Polybius described successive waves of Celtic tribes seeking loot or, more often, territory, across the Alps, and eventually much of the Po Valley in north Italy was overrun (Rankin 1996, 103-15). These incursions need to be considered in reference to the subsequent events in southern France on three counts: firstly, the weakening of Etruscan power in the north aided Rome's eventual emergence as the state power of Italy, and secondly, the memory of the sacking of their city, was remembered and mythologized long after the event (Freeman 2002, 7), thus colouring the perceptions and tone of subsequent Roman histories.

The Siege of Massalia c390 BC

A third outcome of the attack on Rome is a further insight into the relationship between Rome and Massalia. In an event recorded only by Pompeius (Justinus, 43.5):

The Greeks of Massalia made great wars against [their neighbours] the Ligurians and Gauls...when Massalia was at the zenith of its renown...the neighbouring tribes unexpectedly conspired to lay waste to Massalia utterly, so as to wipe out its very name, as if to quench a conflagration that threatened to consume them all

After successfully riding this storm, the Massaliotes sent envoys of thanksgiving to the sanctuary of Apollo at Delphi. On discovering the plight of Rome, the delegation returned to Massalia where a subscription was collected from both private and public funds, to be sent to Rome to help pay the Gaul's ransom. In return, Rome granted Massalia special privileges, including an exemption from custom duty. Ebel believes that such glimpses of the long term rapport between these two cities may go some way to explaining the restraint with which Rome by-passed southern Gaul throughout her expansion during the second century BC. Unlike Sicily, Sardinia, the Spanish provinces and Africa, the documentary evidence suggests that Mediterranean France underwent a relatively tolerant, even incidental, Romanization (1976, 11-12). A perspective which is, to some extent, contested by the thorough degrees of destruction perpetrated by Roman forces at native sites such as Entremont and Roquepertuse in the late-second century.

Third century BC: War in the West

The Attack on Delphi and Migrations into Asia Minor

The third century BC saw the advent of Latin writings on the Celts, however, in many ways the various histories persisted in the same vein as the preceding century. The raiding appears to have continued. Pausanias (fl. AD 173), for example described the Celtic invasions of Greece, and, in 279 BC the ferocious attack on the sanctuary at Delphi which was finally repelled by a combination of divine intervention and Gaulish drunkenness. In 270 BC Gauls (Galatae) settled near Ankara in Asia Minor, and apparently soon found their own place among the native populations (Justinus *Epitome* 25.2; Strabo *Geography* 4.1.13).

The Battle of Telamon

The final phase of these migrations was signalled by the defeat of Celts in north Italy at the Battle of Telamon in 225 BC. Polybius described in detail how a vast Gaulish army became trapped between two Roman forces (*Historia* 2.27. 3-10).

About forty thousand Celts were slain and at least ten thousand taken prisoner...thus were destroyed these Celts during whose invasion, the most serious that had ever occurred, all the Italians and especially the Romans had been exposed to great and terrible peril (2.31).

In the following two years Rome not only drove the invaders into north Italy, successive consuls crossed into the Insubres own territory and, along with their Gaulish allies, laid the country to waste. Although not directly involved in the conflict, the account seems to give an indication of a fairly amicable relationship between at least some of the Celtic tribes around Massalia and Rome (2.32):

Next year's Consuls, Publius Furius and Gaius Flaminius, again invaded the Celtic territory, through the country of the Anares who dwelt not far from Marseilles. Having admitted this tribe to their friendship, they crossed into the territory of the Insubres, near the junction of the Po and Adda

Rome and Carthage

Polybius saw this extended struggle with the invading Celtic tribes of the migration period as a time when Rome emerged perforce as a military power to be reckoned with (2.20, 8-10). In the course of the third century, Rome had already taken control of Sicily and Sardinia in 241 BC and 238 BC respectively (Ebel 1976, 2). However, Walbank argues that a repeatedly demonstrated hesitancy to engage in battles and inconsistency of purpose in Roman foreign policy suggests that, up to the end of the third century, Rome did not hold any concrete aspirations towards imperialism in the west (1972, 164-65). This strategy appears to have changed throughout the course of the Second Punic War, a conflict which had repercussions for both the Phocaean and indigenous peoples of southern France.

Rome, Carthage and Southern France

The Second Punic War (219-202 BC) became the pivotal episode in Rome's past around which many Latin writers shaped histories and epic poetry. According to Polybius, in the course of the First Punic War (264-241 BC) Rome expelled the Carthaginians from Sicily and took control of Sardinia and Corsica. After the war, crippling reparations were demanded, apparently designed to keep Carthage in check, and Punic forces were prohibited from crossing the river Ebro (1.62.9 and 3.10.1-3). The subsequent decades saw a resurgence of Carthaginian power at a time when Rome was occupied with the Gallic invasions of Italy (2.13). The catalyst for the outbreak of new hostilities came when the Carthaginian leader Hannibal captured the Saguntum, a Spanish city with Roman allegiance, and crossed the Ebro (Livy 21.18). The earliest extant sources for the Second Punic War come from the histories of Rome by Livy (Book 21) and Polybius (Book 3). There is also an account of the war, *Punica* by Silius Italicus (AD 25-101).

The Allies and Enemies of Rome in Southern France

According to Livy (21.20), after the fall of Saguntum, the declaration of war with Carthage was discussed in the senate and it was decided to rally support for the Roman cause in Spain and Gaul. The reception of the Roman envoys by the Volciani was repeated throughout Spain. None of the tribes wished to incur the enmity of Carthage, and the Romans were berated for not coming to the aid of Saguntum sooner. The envoys then passed into Gaul:

Much the same was said and heard in all the other assemblies throughout Gaul…not a friendly or even tolerably peaceful answer was received until they came to Massalia. There they got all the information which their allies had carefully and faithfully acquired for them; that already Hannibal had gained a hold on the minds of the Gauls

According to most sources, Hannibal ravaged the lands of the Volcae west of the Rhône. Polybius and Silius both went on to describe how the Volcae confronted Hannibal's force on the eastern side of the Rhône and were quickly defeated and scattered (Polybius 3.42.4). When the Roman general Publius (Scipio) arrived in the region he assumed that the Carthaginian army would still be negotiating the Pyrenees, until his allies at Massalia informed him that Hannibal was already encamped at the Rhône. Scipio then sent out Roman cavalry along with some Celtic mercenaries who had been in the service of Massalia (Polybius 3.41. 5-9). [Scipio] sailed round the Pyrenees and brought his ships to anchor at Emporiae, [from there]…he brought under Roman dominion the entire coast as far as the river Ebro (Livy 21.60). Scipio appears to have used Emporia as a base for his campaign, although confusingly, in Silius' epic poem, the settlement was included in the long list of Hannibal's allies: Emporiae, colony of Massalia (*Punica* III, 365-70). This may have been necessary poetic license as both Polybius and Livy rarely use place names in Gaul. Ebel believes, Hannibal's relatively easy passage through Languedoc and the sudden arrest of his march at the Rhône, illustrates that Massalia's sphere of influence stopped at the river (1976, 35).

Hannibal crossed the Rhône taking horses and elephants over on rafts, while some of the men waded across with their weapons held above their heads: [Hannibal] moved on through the territory of the Tricastini, and made an easy march through the land of the Vocontii (*Punica* III, 455-70). On the far bank Hannibal was met by a delegation of Boii Gauls from Cisalpine Gaul who warned him of the arrival of Scipio in southern France, and offered to guide his force over the Alps and into Italy (Polybius 3.44.5) (Livy 21.29.6). When Hannibal heard that a Roman fleet was stationed in the mouth of the Rhône he sent out his Numidian cavalry to investigate. These horsemen clashed with the soldiers and Gallic guides sent out from Massalia, and, although outnumbered, the Roman and Gaulish detachment defeated the Numidians. This persuaded Hannibal to turn his troops north and he led his army over the Alps and out of France (Polybius 3.45.2) (Livy 21.29.2-3). Hannibal went on to ravage Italy and Sicily for another fifteen years before being finally defeated by Scipio.

Second Century BC: The Roman Expansion

The histories of the second century BC record a time of rapid Roman expansion, not only along the Mediterranean, but south into Africa and northward into large areas of France. After the Second Punic War, with a permanent foothold in Spain, Rome secured her influence in southern France along the path to this new province, but obvious signs of territoriality, such as the construction of a road to Spain, appear to have been restricted to Languedoc. This suggests that the relationship between Rome and Massalia had been strengthened by their allegiance against Hannibal, and that consequently, throughout the period of expansion, Massalia's sphere of influence east of the Rhône was respected (Ebel 1976, 3). The archaeological evidence does concur with a continuation of predominantly Greek, rather than Roman cultural influence (*ibid.*, 104), however, during this century there is a definite awakening of Roman economic interest in southern France centred on the Rhône valley as a profitable conduit into central Europe. Concentrations of Dressel I wine amphorae dating from the middle of the second century have been found all along the southern French coast and up the Rhône corridor (Cunliffe 1997, 312).

Rome's forbearance regarding Provence did not preclude repeated Roman military action in the area which finally culminated in the annexation of the province in the last quarter of the century. Wars in defence of her own territories spilled over into the region but more often troops were sent to the aid of Massalia, whose troubled relationship with the local populations came to a head during this time. In 181 BC Massalia sought Roman aid against Ligurians pirates (Livy 40.18.4-8). Rome had been in conflict with Ligurians in Italy at least since the Battle of Telamon in 225 BC, and it has been postulated that Massalia's cry for help was a diplomatic reluctance to overstep her authority by engaging in battle in Roman waters (Ebel 1976, 61). Whatever the reasons, this was the first significant intervention by Rome in the political affairs of Transalpine Gaul.

The first campaign to result in a brief but significant occupation in southern France was recorded by Livy (Per. 47) and Polybius (*Historia* 33.8). According to Polybius, in 154 BC messengers were sent to Rome asking for assistance in the defence of two Massaliote colonies, Nicaea (modern Nice) and nearby Antipolis (Antibes), once more, against the Ligurians. The Roman delegation sent to investigate the situation was attacked by the Oxybii, one of the Ligurian tribes, and a senator was injured and some of the emissaries killed. Consequently the senate sent Quintus Optimus against the Oxybii and their allies the Deciatae, and the Ligurians were defeated near Nicaea. The tribes were forced to surrender their territories to Massalia and Roman troops spent their first winter in southern Gaul among the scattered Ligurian tribes. The following year another Ligurian tribe, the Lusitanians, staged a short-lived but rapidly expanding insurrection which was promptly put down by the Roman army (Appian Iber. 56).

125-24 BC M. Fulius Flaccus's Campaign

The 154 uprising prefigured the more lasting consequences of the series of campaigns staged in the south of France between 125 and 121 BC involving Roman forces, the citizens of Massalia and the indigenous tribes around Provence. The archaeological evidence described below is suggestive of an increase in native organisation in the second century as well as some of the scenes of combat. Specific references to battles and dates are rare, but the source of the conflict would appear to lie in the growing centralisation among the Gauls, accompanied by the emergence of federations among Gallic and Ligurian tribes in as far as the different 'peoples' or ethnicities were understood by the contemporary chroniclers or extrapolated by modern scholars. The Saluvii, a 'Celto-Ligurian' federation whose capital was at Entremont (see Chapter 3), had emerged at the beginning of the century as a growing power and by 125, they apparently felt it within their ability to challenge Massalia.

C. Sextius Calvinus's Campaign

According to Livy (*Per.* 61) first the Roman consul of 125, M. Fulvius Flaccus, and subsequently the consul of the following year, C. Sextius Calvinus, were sent to the aid of Rome's old ally. While M. F. Flaccus returned to Rome after his campaigns, Strabo (*Geographica* VI.15), reports that, after forcing the Gauls to retreat from the coast, their territory was handed over to the authority of Massalia, and Sextius established a city and barrack near Entremont at the site of a spring: Aquae Sextae (modern Aix-en-Provence), to protect the route through southern France to Rome's Spanish provinces.

Cn. Domitius Ahenobarbus's Campaign

Although Sextius had defeated the Saluvian forces, at least two of the Federation's leaders had escaped, and fled to the powerful Gallic tribe of the Allobroges, for protection. Cn. Domitius Ahenobarbus, was sent, not only on behalf of Massalia, but on the request of a Gaulish ally, the Aedui. The battle between Domitius' forces and the Allobroges, took place at Vindalium. While details of all the campaigns of the gradually increasing Roman intervention in southern France are vague, the clash between Rome and the Allobroges appears to have been momentous. The Arverni tribe came to the aid of the Allobroges, and in 121 BC reinforcements under the command of Q. Fabius Maximus were summoned from Rome. Most of the sources agree that after their defeat, the Gaulish death toll amounted to around 200,000 (e.g. Livy Per 61; Strabo 4.1.11 and 4.2.3).

Caepio and Marius

After the campaigns of 125-21, a series of garrisons were established through the region, principally Narbo Martius (modern Narbonne) in 118 BC (Ebel 1976, 77). It was at Narbo that a further large scale conflict took place in 106 BC when Caepio was sent from Rome to put down trouble amongst the Volcae and Tectosages. Caepio pillaged the town of Tolosa and absconded with its treasures which had been placed as votive deposits in nearby lakes. The following year the Cimbri, a tribe from the north, defeated him and he was exiled, all his misfortunes apparently rooted in his plundering of the sacred treasures (Strabo 4.1.13; Justinus 3.3.36). In 102 BC Marius was sent to deal with the Cimbri and the Teutons as their migrations brought them within the Province. Marius defeated both tribes near Aix.

The date of the foundation of the Roman province of Transalpine Gaul is commonly taken to be around 122-21 BC, initiated by the campaigns of 125 BC and culminating in the defeat of the Allobroges. Nevertheless, as has been remarked above, the sources for this century are vague, and details of the political and military actions relating to the south of France are rare. There is no evidence for a lex provinciae, the necessary legislation for annexation, though of course this is not proof that the process was not instigated. Sicily for example, is known to have a lex which post-dates by some decades its Roman administrative organisation. Perhaps the best that can be securely stated is that after the Cimbric wars regular governorships from Rome commenced in southeast France. Ebel argues that while Roman organisation may have started at some time prior to this, the Romanisation-proper of Transalpine Gaul can definitely be said to begin at this time (1976, 77).

First century BC

58-51 BC Caesar's Gallic Wars

The Gallic Wars began while Caesar was proconsul of Cisalpine Gaul, Transalpine Gaul and Illyricum (Ebel 1976, 105), and Caesar's *De Bello Gallico* is the major source on the conflict. Hostilities in the province are confined to Book I, and the high degree of native co-operation gives some indication of the extent of Roman 'pacification' and control in the region. The war was precipitated when the Helvetii, a powerful and 'warlike' Gallic tribe living in a restricted land on the shores of Lake Geneva, conspired to invade the rest of Gaul with their allies the Boii. Their leader Orgetorix attempted to engage the aid of the southern French Allobroges (I. VI):

There were in all two routes, by which they could go forth from their country one through the Sequani…the other, through our Province, much easier and freer from obstacles, because the Rhône flows between the boundaries of the Helvetii and those of the Allobroges, who had lately been subdued, and is in some places crossed by a ford. The furthest town of the Allobroges, and the nearest to the territories of the Helvetii, is

Geneva. From this town a bridge extends to the Helvetii. They thought that they should either persuade the Allobroges, because they did not seem as yet well-affected toward the Roman people, or compel them by force to allow them to pass through their territories

Hearing of the intended invasion Caesar reported that he gathered his legions and conscripts from all parts of southern France and built a fortification across the intended line of the Helvetii march, informing a delegation of Gallic ambassadors that in accordance with Roman custom, he could not allow the passage through their province (I.VIII):

The Helvetii…tried if they could force a passage (some by means of a bridge of boats and numerous rafts constructed for the purpose; others, by the fords of the Rhône…but being kept at bay by the strength of our works, and by the concourse of the soldiers, and by the missiles, they desisted from this attempt.

Caesar successfully turned the Helvetii back from Provence, and routed the invaders to the Saône at the request of the Aedui and the Allobroges, who claimed the armies of Orgetorix were devastating their lands (I.XI). Caesar went on over the following seven years to conquer much of Gaul and made exploratory but temporary visits to Britain. This episode, however, offers a stark contrast to the previous century's seemingly non-stop outbreaks of rebellion and invasion and indicates the extent to which Roman administration had permeated the region.

The Roman Civil War

Roman politics throughout this time had experienced a series of disruptions resulting from changes in political organisation and internal power-struggles. In 83 BC Sulla had taken command of Italy. In the course of this coup, Rome's eastern interests were secure, however, the transfer of power did not go so smoothly in the west. Resistance to the new order centred to a large degree around Sertorius, who, from his base in Spain, led many successful attacks against Sulla's forces. Opposition to Sertorius was entrusted by the Senate to Pompey in 77 BC. Although technically his commission sent him to Spain, Pompey was obliged to put down a series of uprisings throughout southern Gaul, before making the region the headquarters of his campaign, (Ebel 1976, 96-97). In his account of the civil war Caesar says that at this time Pompey took lands from the Helvetii and the Volcae, and granted them to Massalia (1.35.4). Ebel suggests that this indicates that the relationship between Rome and Massalia was still amicable, and that the Greeks still retained much of her independence and territory throughout the period of Roman administration (1976, 99).

For twenty years, sporadic uprisings in Gaul, for example, the Allobrogic rebellion in 66 BC, resulted in an oppressive form of administration in southern France (Jullian III, 120-21). At the same time, loyalty was bought by gifts of citizenship to the native tribes who gave Rome their allegiance during the struggle. Ebel postulates that by the time Caesar took over administration of the region in 59 BC, this vastly increased Romanisation must have brought the autonomy of Massalia into contrast with the rest of southern Gaul (Ebel 1976, 101). This autonomy finally came to an end ten years later, when the growing rivalry between the two commanders, Caesar and Pompey, erupted into civil war. Domitius, a long-time political enemy of Caesar's, took advantage of the split between Caesar and the senate to secure the position of proconsul in Transalpine Gaul. In the ensuing clash between Caesar and Domitius, Domitius fled and took refuge at Massalia (*ibid* 84-86). Lucan describes the consequent siege of Massalia in his poem about the civil war (*Pharsalia* Book III) and has Caesar bring down the Greek city in a bloodthirsty, thrilling and completely fictitious sea battle (*Pharsalia* III 705-715):

Tyrrhenus was standing on the lofty bow of his ship, when Lygdamus, a wielder of the Balearic thong, aimed a bullet and slung it; and the solid lead crushed his hollow temples. The blood burst all the ligaments, and the eyes, forced from their sockets, rushed forth (etc.)

According to more pedestrian and reliable accounts (e.g. Strabo IV. 1.5), Massalia surrendered under threat of starvation, thus signalling the end of Greek independence in the region. From this point onward, the various populations of southern France were subjects of Rome (Ebel 1976, 105).

4.3 Conclusion

Despite the convivial overtones of the Protis myth, the advent of the Greek-Phocaean colonies in southern France and the foundation of Massalia appear to have resulted in clashes with the indigenous populations from the start. Unfortunately this friction is merely alluded to in a general way in reference to the first three centuries of colonisation. Justinus's mention of the siege of Massalia in 390 BC is one of the few distinct events which can be gleaned, and this is only an aside in the account of the sack of Rome. The first conflict in southern France which is recorded in detail relates to the first Roman struggles to be played out in the region. The impact of the Carthaginian army in southern France at the end of the third century was largely confined to Languedoc, however, it gives a useful indication that Massalia's influence appeared to be centred east of the Rhône. It can also be postulated from their part in this war, as from their offer of aid to their Italian allies after the sack of Rome, that they held considerable power at the time.

The second century BC seems to imply a change in the balance of power in southern France. The repeated episodes of Roman military action in the region appear to be in response to requests for aid by Massalia against her Ligurian and Gallic neighbours. Although this might indicate a decline in Greek strength, it equally appears to highlight increased centralisation among the Gauls, which was strengthened by alliances among Gallic and Ligurian tribes. The 'Celto-Ligurian' Saluvian federation which

challenged Massalia in 125 BC required the successive campaigns of M. Fulius Flaccus and C. Sextius Calvinus to be suppressed. Perhaps the most significant indication of this trend of native organisation and centralisation appears in the descriptions of the battle with the Allobroges in 121. Quite apart from the scale of the conflict, the accounts lack the usual references to 'barbarian' disorganisation, and on occasion approach something like respect for an equally placed enemy. The defeat of the Allobroges, however, signalled an almost complete dissolution of this growing native cohesion, and the apparently region-wide compliance with Caesar's confrontation with the Helvetii in the mid-first century illustrates this point. When Massalia took Pompey's side in the civil war in 49 BC and lost her autonomy, it appears that the last vestige of non-Romanised society disappeared from the area. According to the historic sources, the region finally demonstrated what Caesar termed:

the civilization and refinement of [our] Province (I.I), in contrast to the alien nature of societies beyond the Alps.

4.4 Evaluation of the Historic Sources

Survival of the Original Texts

A full evaluation of all the sources quoted above is beyond the parameters of this study; however, it is necessary to include a short review outlining the principal limitations of the historical texts, with emphasis on the most prominent authors. The reliability of the histories referred to above is weakened by the fact that the sources covering the first four hundred or so years are not extant. Hecataeus, Pompeius and the other early authors only survive through copyists, and are often much abridged (Koch 2000, 5; Collis, 16 and 24). Prior to the sack of Rome in 390 BC, records appear to have been rare, and according to Briscoe (1971), most of these were subsequently destroyed in the attack. Surviving texts probably consisted of family annals which described familial disputes and wars; and some consular lists. After the sack of Rome, the main events of each year were recorded in the *Annales Maximi* (1971, 5). Fragments of Q. Fabius Pictor (b254 BC) 'the first Roman historian' (Collis 2003, 18), survive, but he is usually regarded as a source for later writers (*ibid*); Walbank, for example, identifies Fabius as one of Polybius's main sources for the Gallic Wars and the First Punic War (1972, 106). Polybius (c. 205-123 BC) was the earliest prominent author whose work survives to a significant degree. Substantial sections of the histories of Livy (64/59 BC-AD 17) are also available, and together, they constitute the earliest and most important extant sources on this period (Proctor 1971, 6).

4.5 The Archaeological Confirmation?

4.5.1 The Greek Settlement

Punic and Etruscan, as well as Greek activity has long been acknowledged in Mediterranean France from the seventh century BC; however, this has been interpreted as, at most, a series of entrepôts or trading posts rather than colonisation. Traditionally Marseilles has been considered the 'earliest definite Mediterranean settlement on Gaulish soil' (Py 1993, 45). This opinion has recently been reviewed, and current excavations now appear to indicate permanent and substantial Etruscan foundations, (such as Lattes) particularly in Languedoc (Thierry Janin pers. comm.). As mentioned above, Provence appears to have constituted the main sphere of Massaliote influence. The foundation of the city and her daughter colonies in southern France by Phocaean Greeks are borne out by the archaeology of the region. Despite a long and continuous sequence of occupation, the remains of domestic structures and ceramic objects have survived in Marseilles, as well as a large underground conduit for bringing water into the city; all of which date to the beginning of the sixth century (Moliner 1990, 42-43).

Finding archaeological evidence for specific events, in particular the Gaulish and Roman sieges of Massalia in 390 BC and 49 BC, is more problematic. Trousset suggests that the construction of large water cisterns in Marseilles can be seen as a defensive measure. He refers to Lucan's account of the siege of 49 BC, when Caesar cut off the Massaliote from their external resources by excavating a ditch outside the city. While rejecting Lucan as a reliable source, he proposes that the depiction might reflect an essential response to the dangers of siege warfare, alluding to an injunction by Aristotle that each [Greek] city should ensure it had its own water supply in case of war (*Politique* VII, 10) (1993, 38-39). The most explicit evidence relating to this study appears to come from the south tower of the city's rampart, which was discovered with the remains of catapult installations, used, according to Salviat, to defend the city against Caesar's army in 49 BC (1993, 29).

4.6 Conclusion

With the exception of the sole mention by Pompeius of the siege of Massalia, there appear to be no unequivocal reports of specific wars in the southern French Iron Age

earlier than the versions of the Second Punic War recounted by Livy and Strabo. All three of the above authors were Roman and all three wrote in the first century BC, so that these accounts must be seen as coming from a militaristic and newly-colonial perspective. Roman interest in southern France was awakened during the Second Punic War and, around twenty years later (181 BC), in Provence, with the appeals from Massalia for aid against Ligurian pirates. Prior to this, descriptions of violence are vague references to the indigenous populations as being 'warlike' and savage. Webster (1996) objects to the adoption of such ideas of 'Celtic warrior societies' by archaeologists, and argues that the branding of Gauls as 'barbarian' by Classical writers has resulted in an erroneous picture of a people involved in recurrent internecine fighting; whereas a great deal of the conflict recorded in the Late Iron Age consisted of hostilities directed against Rome, rather than inter-tribal warfare. She also notes that such a construct lacks any 'specific historical context', merely adding to the myth of 'the timeless primitive' (1996, 111).

Without casually accepting the idea that the indigenous populations of Provence were innately warlike because they were 'barbarian' or 'Celtic', the existence of a colonial 'tribal-zone', (whether Greek or Roman), would probably have created the stimuli for internal warfare. This situation has been recorded many times in modern ethnographies, for example the case of the Iban tribes of southeast Asia mentioned in chapter two, whose internal patterns of warfare and headhunting drastically intensified after contact with the state-level societies of Britain and China (Gibson 1990). In the case of southern France, the lack of historical context would merely reflect the natural inclination for Roman authors to record only the hostilities in which they themselves were engaged. The extent to which Greek and Roman authors were simply recording a mythical cultural 'other' who engaged in warfare as part of a set of 'barbarian' practices including drunkenness, homosexuality etc. is also a factor which impacts on this supposition, and, therefore, is discussed briefly below.

SITE	LOCATION	FOUNDATION DATE
AMPURIAS (SETTLEMENT)	SPAIN	AFTER 600 BC (BECAME INDEPENDENT OF MASSALIA IN THE 5TH CENTURY BC)
AGDE (SETTLEMENT)	HÉRAULT	END OF 5TH CENTURY BC
ESPEYRAN (ENTREPOT)	GARD	6TH CENTURY BC
ARLE (PORT / ENTREPOT)	BOUCHES-DU-RHÔNE	C. 350 BC
TAUROEIS (ENTREPOT)	VAR	END OF 3RD - START OF 2ND CENTURY BC
OLBIA (SETTLEMENT)	VAR	C. 330 BC
ANTIBES (SETTLEMENT)	ALPES-MARITIMES	
NICE (SETTLEMENT)	ALPES-MARITIMES	

Table 4.2 Greek colonies and trade centres

4.7 Part II: The Ethnographic Accounts

4.8 Introduction

The histories of southern France discussed above have concentrated on events transcribed between the foundation of Massalia to the beginning of Caesar's *Gallic Wars*. Much of the documentary evidence regarding the region, however, pertains to the character and customs of the indigenous peoples, giving some account of their practices and reflecting something of the interaction between the native populations and their Greek and Roman colonisers. As mentioned above, the earliest sources were Greek, and derived from centuries of maritime trade and settlement. Seafarers required some knowledge of, not only the coastlines of their journey, but the harbours and settlements along the way, including the friendliness or otherwise of the inhabitants. Thus the early Greek writings present a blend of periplus, geography and ethnography (Tierney 1960, 189). Roman sources began later; and the end of the first century BC even saw the emergence of indigenous writings from the early Empire, although, of course all the authors mentioned in this chapter were heavily 'Romanised' and wrote in either Greek or Latin. Pompeius (late first century BC) for example was a member of the Vocontii tribe of southern France, Lucan (first century AD) was Spanish and another first century AD poet, Martial, claimed Celtiberian decent (Collis 2003, 22-23). One of the most important ethnographers of the ancient Gauls, however, was a Syrian Greek, Poseidonius, whose works exist now only through later copyists.

The historical and philosophical background of each author informed the approach of their various narratives. The concepts of 'hard primitivism' and 'soft primitivism', discussed more fully by Lovejoy and Boas (1935), are summarised by Piggott as reflecting the degree of contact with the subject of the narrative. Distance, either temporal or geographical, allowed a certain softening of perceptions, so that 'primitive' customs became quaint, or at least interesting rather than repellent. So, for example, Herodotus (4.93) could refer to the Getai in the distant north as 'a most courageous and most just people' (Rankin 1996, 52). Conversely Aristotle (384 - 322 BC), who was writing at a time when the Celtic migrations and the assaults on Rome and Delphi were still raw and recent events, describes Celts as being reckless in a manner that denotes madness rather than bravery (*Ethic*. Nicomach. III, 7.7). Poseidonius' ethnography of the ancient Gauls appears to have achieved a position somewhere between the two.

4.9 Evaluation of the Ethnographic Sources

Poseidonius

Poseidonius (c 135 - 51 BC) wrote his *Histories* around 80 BC (Nash 1978, 111) over forty years after the establishment of a permanent Roman military presence in the region. Although the extent of his travels in unclear, it is believed that he visited parts of Gaul and spent some time at Massalia, where he would have drawn on both written and eyewitness accounts (Piggott 1968, 83). His proximity to the indigenous population should bring Poseidonius into the realm of 'hard primitivism', however, his adherence to the precepts of Stoicism rendered him, as Piggott describes 'a hard primitivist with a rather soft centre' (*ibid*, 77). His copyists, Athenaeus and Strabo, were influenced by the same philosophy, partially set out here by Tierney (1960, 212) as recounted by Seneca:

The barbarian nations…in their customs and civilisation represent a stage left far behind by the culturally advanced peoples, often in their simplicity and virtue recall the psychology of the Homeric or even the Golden Age. Even the barbarians, however, have degenerated and we can see among them both superstition and tyranny (Seneca Ep. 90).

Poseidonius drew on many earlier sources, some of whom, Aristotle and Polybius for example, may have had first-hand knowledge of the Gauls. The *Histories* was to be a continuation of Polybius's own history of Rome, beginning where Polybius finished, in the middle of the second century BC, consequently his ethnography of the Celts (Book 23) starts with the account of the Roman military campaigns from around 125 BC, and the establishment of Gallia Transalpina. Although no longer extant, the writings of Poseidonius are considered to form the basis of much of the important surviving ethnographic accounts on the Celts. The extent of his influence has been the subject of some debate (see Jacoby n.d.; Tierney 1960; Nash 1978) and while this study is more concerned with the reliability of these accounts and the inter-cultural relationships reflected in them, the arguments are presented, very much simplified, below.

Tierney (1960) attempted to reconstruct the original source-material of Poseidonius by comparing similarities and omissions in his four best-known copyists, Diodorus Siculus, Caesar, Strabo and Athenaeus. He also refers in these texts to the kinds of literary convention which would have been employed by Poseidonius and other Greek ethnographers, such as Herodotus. Nash (1978) feels that Tierney's approach includes several drawbacks, such as the likelihood that subsequent writers may have copied from each other as well as from Poseidonius's own sources. She also proposes that some authors, particularly Caesar, probably drew on their own experience of Gaulish life, and that divergences with other accounts should not be dismissed as an inferior facsimile of Poseidonius.

Diodorus Siculus (fl. 58 BC)

Diodorus's *Bibliotheca Historica* was intended to be a world history in digest form. A compilation rather than an original work, the Bibliotheca drew on numerous sources including Ephorus, Timaeus, Hieronymus and Polybius, and provides a unique access to many other sources which are no longer extant (Stylianou 1998, 1-2). A contemporary of Poseidonius, Diodorus's Chapters 23 -

32 in book V are believed to derive from his history as are the corresponding chapters in Strabo (Tierney 1960, 203). Regrettably it is generally agreed that the *Bibliotheca Historica* is a very bad compilation: authors are misrepresented and expanded with pretentious verbiage (Stylianou 1998, 132) or arbitrarily and clumsily condensed. Poseidonius is very much abridged in Diodorus, especially regarding the geography of Gaul. His passages concerning the people of Gaul suffer less from these cuts but Strabo presents a more complete version of the same material (Tierney 1960, 204). Diodorus intended his publication to act as a source of moral instruction (Stylianou 1998, 3-14), and often his versions of the accounts of Gaulish customs appear to be couched in the harshest tones, his style being, perhaps self-consciously, apposite to the moral propriety promoted by Augustus in the first century BC. Diodorus also had a tendency to generalise, for example he described the Mistral as raging over all of Gaul, whereas Strabo specifies that it occurred in La Crau (Tierney 1960, 204). As discussed below, this last trait may have influenced the collection of standard anecdotes which would eventually form the basis for the characterisation of a 'Celtic temperament'.

Caesar (100 - 44 BC)

Caesar presents some problems, as he does not acknowledge Poseidonius as a source in *De Bello Gallico*. Nash believes that if some of Caesar's descriptions differ from Athenaeus and the others, it could be because he was presenting original information from a different historical context rather than reproducing an inferior copy of Poseidonius. For the purposes of this discussion, the extent of Caesar's use of Poseidonius provides few major problems as most commentators agree that Athenaeus's version is the most complete and faithful facsimile of the original, even coming from the same school of Stoic philosophy (Tierney 1960, 201; Nash 1978, 112); and that Caesar, while drawing on Poseidonius, should also be viewed as an original source in his own right.

Caesar's brief descriptions of southern France present an image of a perfectly 'Romanised' society very different to the picture of Celtic life apparently composed by Poseidonius seventy years before. The perfect cultural syncretism described in the first chapter of *De Bello Gallico* is questionable. Caesar was writing with a political agenda, and in his justification of his war against the Suebi he was specifically assuming the role of Roman 'pacifier' of the barbarians and defender of the 'civilised' Province of Transalpine Gaul, thereby building a personal reputation for the advancement of his political career (Collis 2003, 20). Caesar was further constrained by a need not to deviate very far in his accounts from his original reports to the Senate (Tierney 1960, 212). One of the most frequent allegations against the validity of Caesar's few ethnographic passages concerns his emphasis of the pan-tribal power of the druids in Gaulish society. It may have suited Caesar's purposes to inflate

the danger of centralised power uniting the Celts, and, indeed, the political role of the druids is not emphasised either by Athenaeus and Strabo (Tierney 1960, 214-15). Tierney suggests that the political power of the druids may also have been exaggerated by Poseidonius, as the stoics believed that the philosophers of the Golden Age ruled both politics and religion (*ibid*). However, Nash believes that the lack of evidence concerning the druids outside such accounts makes a dismissal of either Poseidonius or Caesar unwise (1976, 124).

Strabo (c. 63 BC - AD 21)

Strabo is believed to have utilised Poseidonius extensively in Book IV Chapters 1- 4 of his Geographica (Tierney 1960, 207). Strabo was a near contemporary of Poseidonius, and although he was acquainted with some of the most famous philosophers and intellectuals of his day (Dueck 2000, 8-15) suggestions that the two men may have met (Piggott 1968, 83) appear to be a fallacy based on a misreading of Strabo's work by Athenaeus (Diller 1975, 9; Dueck 2000, 10). Nash considers Strabo to be an unsound source for Poseidonius as much of the subject matter in the *Geographica* either post-dates Poseidonius or could just as easily be attributed to earlier sources, Strabo having cited another nine writers on the subject of the Celts (1976, 113). It has been suggested that the Gaulish material in Strabo which was unavailable to Poseidonius, for example the classifications of the peoples of northern Gaul into distinct groups, can probably be attributed to Caesar (Tierney 1960, 207; Dueck 2000, 93). This presumption, however, does appear a little too simplistic considering the breadth of sources (written and living) which Strabo may have used, and Nash's point, for example, that it is virtually impossible to distinguish Poseidonius from Timagenes quoting Poseidonius in the writings of Strabo is compelling (1976, 113).

For present purposes it is mostly important to emphasise that Strabo was familiar with Poseidonius along with most of the main sources available in the first century BC, and that, although well-travelled, there is no evidence that he ever visited Gaul himself (Diller 1975, 3). While open to a variety of influences, Strabo displayed a strong tendency towards stoicism in the *Geographica* (Dueck 2000, 64). Dueck points to his obvious disapproval of Roman luxury and excess, and also his determination to present the more unusual 'marvels' of foreign lands in a calm and self-possessed manner (Dueck 2000, 64). Nevertheless, she also draws attention to Strabo's very un-stoic position that, rather than sharing a common humanity, there existed a civilised-barbarian dichotomy (*ibid*, 75-84), but that the barbarous peoples decreased in savagery when they came into contact with civilisation (Dueck 2000, 79). These two motifs in Strabo's writing, stoicism and a belief in the civilising effects of Roman rule, lead to some contradictory views on the conquests which brought both Roman order and Roman corruption (*ibid.*, 118-19).

Athenaeus (writing c. AD 200)

Athenaeus composed his *Deipnosophistae* or '*Banquet of the Learned*' in the late second or early-third centuries AD (Thompson 2000, 78). The narrative comprised a series of philosophical discussions at a Roman feast covering the scholarship of Greece, ethnography, philology, cultural history etc for the benefit of a Roman host (Jacob 2000, 86; Hopwood 2000, 231). Athenaeus drew on a vast range of literary sources, most of which are referenced and many of which are now lost (Arnott 2000, 41). According to Tierney, it contained four extracts from Poseidonius which conform to a traditional structure and style similar to that employed by Herodotus and other early geographer-historians, leaving the impression of complete pieces of work. Athenaeus's accounts are also much longer than the other versions, both of which factors suggest that he was reproducing Poseidonius verbatim (1960, 201-02). The credentials of Athenaeus as a source for Poseidonius are generally well accepted (*ibid*; Nash 1975, 112), although Pelling warns that Athenaeus tended to utilise one author as a framework for a long passage of text, while drawing on other unacknowledged sources for small supplementary fragments (2000, 175).

Athenaeus was born in Naucratis, a seventh-century BC foundation established by colonists from a variety of eastern Greek cities in the Egyptian Delta which maintained its independent Greek status under Roman rule (Thompson 2000, 77). Unlike Poseidonius and many of the other authors mentioned above, such as Caesar or Livy, Athenaeus was writing when the upheavals of Rome's internal political struggles and growing empire were long passed. His portrayals of foreign customs appear mainly as illustrations of 'otherness' and stock 'barbarian' characteristics rather than expressing specific issues with Gauls or Carthaginians. Athenaeus seems to have evinced similarities to what is known of Poseidonius's outlook, for example in the backward-looking aspect of his writing, which recalls past virtues in Roman, Greek and other 'civilised' cultures (Braund 2000, 3). He also endorses Poseidonius's position on Roman history as an analysis of an individual and political decline from austerity into luxury (*ibid*, 11). Despite his tendency to look back on a (primarily Greek) Golden Age and his frequent lampooning of gluttony and drunkenness, Athenaeus does not conform completely to the early Stoic position of Poseidonius or Strabo, appearing to see luxury as a necessary consequence of Roman peace (Braund 2000, 12).

4.10 Ethnographic Convention?

Repeated copying from a common pool of sources has resulted in a standard set of anecdotes describing the Gaulish Celts, their fierce customs and their natural propensity for warfare. The perspective of each version changed according to the personal bias of the author, which can be surmised to some extent by analysing the writer's historical background and personal philosophy. Ascertaining whether a source was likely to compose a jaundiced or sympathetic rendering of a certain story, however, does not necessarily shed light on the veracity of any or all of these tales. This section considers the degree to which conventions may have distorted original observations; and to what extent the 'Celtic' practices described can be said to apply to the peoples of the southern French Iron Age. Initially it is necessary to identify those accounts which, profess to describe 'Celtic' behaviour, but in reality form part of a standard Greco-Roman repertoire of 'barbarian' mythology.

Concepts of 'Otherness' and the Warlike Celts

Much of the Celtic or Gaulish ethnographies seem to have begun in earnest in the fourth century BC, the period of the Celtic migrations, in which context, naturally enough, many of the accounts record warlike behaviour. The attributes of the marauder, such as ferociousness and drunkenness, became the basis of the Celtic character. In the ensuing centuries the Celts were portrayed as either dangerous savages or guileless children of nature, however, the lack of restraint which underlies these descriptions probably stems from this time. The idea that non-Greek people constituted a cultural 'Other' is believed to have emerged around 500 BC when the Greeks were at war with Persia (Hall 1989). Whatever its origins, it appears clear that by the fourth century BC the barbarian-civilised dichotomy was firmly established, and the concept of the 'Barbarian' was one which Strabo (14.2.28, C661-3) described as being self-consciously employed as part of Greek literary tradition (Dueck 2000, 76). Rankin postulates that Plato's fourth century description of heavy-drinking, warlike peoples: Scythians and Persians, and also the Carthaginians, Celts, Iberians, and Thracians (Laws I.637d-e), was a conventional list, and that increasing contact during this period resulted in the Celts being 'added to the register' during this period (Rankin 1987, 54). Webster has developed this theme in the context of 'cultural otherness' (1996) and suggests that the concept of barbarism as the converse of civilisation was adopted by Roman authors who came into more prolonged contact with the western Gauls than the Greeks had done, firstly at the time of the Punic Wars, and then more directly from the Transalpine Wars beginning around 125 BC. From this period onward, the idea of taming the barbarian Celts became a useful rationale for the Roman imperial expansion into north and west Europe. These representations of the Celts take the form of either direct ethnographic descriptions or illustrative anecdotes and sometimes it is possible to recognise these as topical and perennial mythologies. For example, Caesar's reference to the 'just and warlike Volcae' (*De Bello Gallico* 24) echo the exact sentiments regarding the Cretans by Ephorus or the early Romans by Poseidonius (Tierney1960, 218). Unfortunately, the reuse of certain themes in reference to a number of different peoples may simply indicate analogous customs, for example descriptions of headhunting Scythians by Herodotus and Celts by Poseidonius (Koch 2000, 36-37). The recurrence of certain customs or themes in Greek and Roman literature therefore appears to provide no indication of the reliability of the material.

If parallels between accounts say little about the credibility of certain practices, a better indication of reliability may be found where these descriptions diverge, giving different, and potentially specific information on the southern Gauls, as distinct from other foreigners. One such example can be found in the contrasting reactions to the Roman plea for help against Carthage by the Spanish and the Gauls (both Barbarian and Other) as described by Livy (21.20): The Spanish refused to come to Rome's aid against the Carthaginians and their spokesman rebuked the Roman diplomats gravely for the abandonment of the city of Saguntum. The Gauls of southwest France, however, behaved in a much less dignified way:

Here they witnessed a strange and alarming sight...The people [the Gauls] came armed to the assembly - their national custom. [When the envoys petitioned for an alliance] there was such a burst, it is said, of hooting and laughter, that the magistrates and the elders could hardly quiet the younger men...

[as recounted above, the Massaliotes warned the Romans that Hannibal had already won over the Gauls]

...but that even he would not find the nation sufficiently tractable (so fierce and untameable was its temper) unless he further won the affections of the chiefs with gold, of which the Gaul is intensely greedy.

The dissimilar nature of the two episodes could indicate that it may contain a kernel of authenticity, although they may simply represent two types or grades of barbarian, (the Spanish scoring rather higher than the Gauls).

Most of the Greek and Roman texts probably adhere to some extent to literary convention and cultural bias, and are therefore, never wholly reliable as source material on the customs of the southern Gauls. Furthermore there appears to be no general rule for the recognition of traditional barbarian folklore in the ethnographies dealing with the indigenous peoples of southern France. It is necessary, however, to consider how much, if any, of this material is supported by archaeological evidence, and how this may reflect the practice of war during the Iron Age in southern France, and any obviously recycled material will be indicated as far as possible in the discussions below.

Soft Primitivism: The Social Role of Celtic Violence

The idea of The Golden Age as described by Seneca above, was the cornerstone of Stoic philosophy. Poseidonius's account of the 'customs and manners' of the Celts of southern France was sympathetically transcribed by Athenaeus (4.36), who was writing long after the decline both of Greek power and the Celtic threat:

Poseidonius, the Stoic philosopher collected the customs and manners of many peoples in his Histories, which he wrote according to his own philosophical convictions. He

says that the Celts...dine on this meat in a clean but lion-like manner...

Tierney suggests that accounts of Celtic feasts and assemblies, such as Poseidonius's description of Celtic feasting or Caesar's account of the council meeting of the central Gaulish Celts (*De Bello Gallico* 23) reflect an overtly Homeric convention regarding the importance of hospitality among more 'primitive' peoples. Poseidonius recounted the custom of mock battles at banquets which sometimes escalated into actual bloodshed and death, as well as the more deliberate violence of single combat between warriors over the 'hero's portion' which was another recurring theme in the literary feast (Tierney 1960, 218). Presumably, the message of the convention was that among the barbarians, both hospitality and violence acted as important social functions, which contributed to public understanding of the hierarchy and power-structures.

Hard Primitivism: The Celtic Warrior and Innate Aggression

Piggott characterises hard primitivism as stemming from direct contact (1968, 77), of which the first large-scale examples were violent. The first example, the foundation of Massalia, was presumably instigated by the Greek colonists, and if any records of specific conflicts did exist, they have not survived. Conversely the Celtic raids into Italy and Greece in the fourth century BC constituted a series of sustained invasion into Mediterranean territory, and the conflicts, apocryphal or otherwise, were copiously documented. Naturally the accounts of these incursions were unfavourable to the Celts, with the requisites of the Classical barbarian laid down in Pompeius' account of the raid on the temple at Delphi in Justinus (*Epitome* 24. 6-8):

He [the Celtic leader Brennus] collected together an army of one hundred and fifty thousand infantry and fifteen thousand cavalry and invaded Macedonia himself. While pillaging the Macedonian countryside, he was met by Sosthenes at the head of a Macedonian army. But the Macedonians were few and frightened while the Gauls were numerous and unafraid, so that they had an easy victory. The Macedonians then hid inside their walled cities while Brennus and the invading Gauls looted the entire countryside unopposed.

Brennus's downfall was twofold: firstly, greed and impiety drove him to raid the temple at Delphi: 'Joking foolishly that they [the gods] were generous and ought to share their goods with men' (Epitome 24.6), causing the Greek gods to take a hand and assault the invaders with hailstorms and landslides. Secondly, the perennial barbarian weakness for alcohol prompted the invaders to get drunk on their spoils, rather than drive their advantage home. 'They [the Gauls] were suffering terribly from the wild night of drinking' (Epitome 24.7) This delay allowed the Delphians to regroup and get help from neighbouring cities, and the Gauls were driven from the city. Direct contact with the Celts often took the form of military conflicts and similarly undisciplined behaviour was described by Silius in reference to Hannibal's Gaulish allies during the Second Punic War. For example the killing of the Gaul Crixus by the Roman general Scipio results in the defeat of the Gaulish forces...

Deprived of their leader, the Gauls had recourse to flight; all their confidence and all their valour depended upon a single life (Punica IV, 302-05)

He then goes on to describe the rout of the Gaulish forces with an analogy of hunters harrying wild beasts.

Mago [brother of Hannibal] saw that the ranks of Gaul had turned back and that their first effort had failed (and that people is incapable of a second) (Punica IV, 310).

These characteristics of ferocity, drunkenness, and disorder were subsequently retold by later writers, sometimes as in the tradition of soft primitivism, as impulsiveness and childlike naivety. Strabo who, to some extent, admired what he saw as the hardiness of barbarian life, summed up the Gauls as fighters rather than farmers (4.1.2, C 178). According to Dueck, he considered the 'barbarians' to suffer from two main vices: a lack of socio-political structures and embedded violence (2000, 116). Ultimately, in both soft and hard primitivism the Celtic temperament was defined by a lack of self-control which was the antithesis of the Greek and Roman ideal of civilised conduct and which inevitably resulted in violence.

4.11 Ethnographic Reality?

According to the ethnographies of Greece and Rome, therefore, aggression and lack of restraint comprised an innate characteristic of the Celts. The belief that any specific population may have a natural predilections for violence has already been dismissed in chapter two, however, Webster (1996) feels that the Classical texts have infused Iron Age archaeology with the construct of the Celtic 'warrior elite' much as modern ethnographies, such as Chagnon's work among the Yąnomami, have resulted in the myth of a 'fierce people' of the Amazonian rainforest (1996, 111-12). The significance of some of the specific practices described in these accounts to illustrate this savageness, however, may have some bearing on the search for evidence of warfare in the Bouches du Rhône.

Animism and Woodland Sanctuaries

The lack of emotional restraint and concomitant lack of socio-political structures, which informed the opinion of the Mediterranean societies regarding the Gauls was epitomised by the tradition of Celtic religious practices in natural places, particularly woodland groves. This tradition allowed soft primitivists to foster the myth of the barbarian, living in harmony with nature, uncorrupted by civilisation, while hard primitivists could cultivate the image of a Wildman, what Webster calls the 'timeless primitive' (1996, 111-13). Webster's misgivings about a-

historical classical authors perpetuating Celtic mythologies in modern scholarship are well illustrated by the regularity of the recurrence of Lucan's familiar passage (*Pharsalia* III.400-425):

A grove there was, untouched by men's hands from ancient times...gods were worshiped there with savage rites, the altars were heaped with hideous offerings and every tree was sprinkled with human gore etc.

The passage, written by Lucan in the mid-first century AD of Caesar's siege of Massalia in 49 BC, implies the carrying out of animistic practices in forests and natural places well into the period of Roman provincial government in the first century BC. While changes in ritual practice in the region are too complex to be explored in any detail here, (see chapters six and seven), suffice it to say that Lucan's depiction of native religious structures seems excessively oversimplified. Although natural places, caves, springs etc., and extra-mural sites continued to be significant throughout the Iron Age in indigenous ritual organisation, the increased sedentism of the Second Iron Age saw the settlement become the essential focus of cult centres and ritual practices, evidenced by votive urns, offerings in pits, stelae, plaques incised with lozenges, preserved human remains etc. (e.g. Arcelin et al 1992, 182).

Headhunting

The most notorious warlike custom of the Celts described in the Greek and Roman texts is headhunting. Ritual practices involving human heads can be found throughout Iron Age Europe, but it is particularly evocative of the indigenous ritual monuments and statuary of the Bouches du Rhône. The statuary and skulls from this concentration of sanctuaries is discussed more fully in chapters six and seven, but for the present it should be stated that human heads appear to have been used as ritual objects in sites and peripheral sanctuaries around the Lower Rhône Valley, and that their associated iconography suggests that they were somehow connected with warfare. The earliest specific mention of Celtic head-taking appears to come from the start of the battle of Telamon (Polybius *Historia* 2. 28):

In this action Gaius the Consul fell in the mêlée fighting with desperate courage and his head was brought to the Celtic kings

The best-known and most comprehensive passage, however, comes from Poseidonius, cited here by Strabo (4.4.1):

In addition to their witlessness, there is also among them the barbaric and highly unusual custom (practiced most of all by the northern tribes) of hanging the heads of their enemies from the necks of their horses when departing from battle, and nailing the spectacle to the doorways of their homes upon returning. Indeed Poseidonius says that he saw this himself in many places, and that while he was unaccustomed to it at first, he could later endure it calmly due to his frequent contact with it. The heads of those enemies that were held in high esteem they would embalm in cedar oil and display them to their guests, and they would not think of having them ransomed even for an equal weight of gold.

The extent to which Poseidonius was writing from actual experience, rather than from documents and informants at Massalia, has been the matter of some debate (e.g. Nash 1976; Webster 1996), however, this excerpt suggests that Poseidonius had personally witnessed displayed heads in the area around Marseilles. The passage also indicates that the practice was extremely common, a supposition which appears to be supported by the skulls and skull fragments from native sites in the Lower Rhône Valley, some still retaining large iron nails.

Sacrifice of Prisoners of War

Strabo and Diodorus described the sacrifice of criminals by shooting with arrows (Strabo 4.4.5) or impaling (Strabo 4.4.5; Diodorus 5.32). Both also recount the sacrifice of numbers of people and animals by burning, either on a pyre or in a large figure of wood and straw (*ibid*). In these accounts Strabo was referring to the Gauls as a whole, while Diodorus was specifically writing about the Cimbri, who, along with the Teutons, were conquered by Marius in 102 BC near Aix and whom he described as evincing more than the usual barbarian savagery. Of the three best-known accounts of these Celtic human sacrifices (Strabo and Diodorus and Caesar) only Diodorus describes these victims as prisoners of war (5.32):

They use war prisoners as sacrificial victims to honour their gods. Some even sacrifice the animals captured in war in addition to the human beings, or burn them in a pyre or kill them through some other means of torture.
There is also a brief, early allusion from Sopater (third century BC) which survives in Athenaeus (*Deipnosophistae* 15.16oe):

Among them [the Celts] is the custom, whenever they win victory in battle, to sacrifice their prisoners to the gods.

It is likely that both Strabo and Diodorus based their versions on Caesar's account below (*De Bello Gallico* 6.16):

All the people of Gaul are completely devoted to religion, and for this reason those who are greatly affected by diseases and in the dangers of battle either sacrifice human victims or vow to do so using the Druids as administrators...since it is judged that unless for a man's life, a man's life is given back, the immortal gods cannot be placated. In public affairs they have instituted the same kind of sacrifice. Others have effigies of great size interwoven with twigs, the limbs of which are filled up with living people which are set on fire from below, and the people are deprived of life surrounded by flames. It is judged that the punishment of those who participated in theft or brigandage or other crimes are more pleasing to

the immortal gods; but when the supplies of this kind fail, they even go so low as to inflict punishment on the innocent.

It is probable that Caesar was alluding to areas of Gaul beyond the Province, which he considered to be largely Romanised at the time of his campaigns. His reference to disease, war and superstition, however, is interesting in light of the possible correlation between unpredictable disasters, witchcraft and endemic warfare (Ember and Ember 1992) discussed in chapters two and five. If burnt human-sacrifices as described above were practiced in southern France or throughout Gaul, however, the physical traces would be unlikely to survive. Furthermore, there are no other similar accounts besides Caesar and his two probable copyists, leaving these descriptions open to question.

'Warrior Sanctuaries' and the Afterlife

As with the accounts of Celtic headhunting, little effort is made in the Greek and Roman texts to discuss the intricacies of Celtic religious practice; which the iconography and vestiges of ritual paraphernalia suggest may have been a complex and evolving set of beliefs. Caesar refers to the burning of the possessions, slaves and clients of the deceased on funeral pyres as a recently defunct practice (6.16), presumably, among the northern Gauls. Belief in an afterlife, as implied by such a practice, is mentioned by several writers, and Valerius Maximus, writing around AD 35, appears to have been alluding specifically to the Gauls of Provence in his assertion of this belief (2.6.10):

Having completed my discussion of this town [Massalia], an old custom of the Gauls should be mentioned: they lend money repayable in the next world, so firm is their belief in the immortality of the spirit. I would say they are fools, except what these trouser-wearers believe is the same as the doctrine of toga-wearing Pythagoras.

This belief, according to Rankin, was related in classical literature to the alleged ferocity of the Celts in war, as they had no fear of death (1987, 52). A similar conviction would appear to have inspired the custom described by Nicander of Colophon which survived in Tertullian (*De Anima* 57.10):

And it is often alleged because of night-time dreams that the dead truly appear, for the Nasamones receive special oracles by staying at the tombs of their parents, as Heraclides - or Nymphodorus or Herodotus - writes. The Celts also for the same reason spend the night near the tombs of their famous men, as Nicander affirms.

A tradition of ritual practices utilising the carefully remodelled and painted skulls, interpreted by Arcelin, Dedet and Schwaller (1992, 215) as evidence for ancestor worship in the Bouches du Rhône may have dated to the third century BC. Nicander was writing in the second century BC, by which time, these 'warrior sanctuaries' with their modified skulls and portrait-like sculptures of severed heads, were in use in settlements such as Entremont, Glanum, or Roquepertuse.

4.12 The 'Celtic Warrior' Stereotype and Southern France

Arguably the most significant ethnographic material on the Celts was drawn from Poseidonius, who, while sometimes described as having travelled extensively in Western Europe (e.g. Rankin 1987, 75; Collis 2003, 20), may very well not have journeyed beyond southern France (Nash 1976, 120). Collis states that Poseidonius was the source of many of the stock anecdotes regarding the Celts, such as their traditions of feasting, human sacrifice and headhunting (2003, 20); although, as mentioned above, the accounts of human burnt-sacrifice probably originated with Caesar, who may be treated as an independent source (Nash 1976, 121) and was probably describing activities north of the Alps. The stereotypical Celt of classical literature: boastful, violent, and melodramatic (Diodorus 5.31), had its roots in the fourth century BC Greek sources, such as Plato and Aristotle, who witnessed and recorded the invasions of Celts from north and central Gaul. Poseidonius's travels in southern France supplied a series of activities which were grafted onto this early character. The result of this combination was a rather one-dimensional caricature which belies both the physical vestiges of elaborate religious practices, for example the head-cult of the Lower Rhône Valley, and the historic accounts of the final clashes between the Roman forces of Caepio and Marius, and the groups of complex and highly organised people such as the Allobroges.

Religious Practice and the Greco-Roman Relationship with Gaul

The manner in which Gaulish ritual practices were recorded in the Greek and Roman texts reflects the shifting relationship between the Mediterranean peoples and the native populations of southern France. With some notable exceptions (e.g. Diodorus) Piggott's correlation between hard primitivism and direct contact appears broadly accurate (1968, 77). In addition the ethnographic and historic accounts, particularly those from the second century BC, must be viewed as part of the 'colonial discourses' of the expanding Roman Empire (Webster 1996, 111). Rankin also states that the Greeks viewed barbarian societies as an illustration of their own archaic and undeveloped condition, prior to the forming of the city-states. In this way, the Celts may have been seen as reflecting traditions from the Mediterranean's own historic past, such as habitually carrying arms (1987, 49-50). It may be this last factor in particular which informed the choice of subject matter in the ethnography of the Celts, as well as the misinterpretations of those customs recorded.

Religious worship in natural places, for example, is often considered to exemplify the savagery of the Gauls, in juxtaposition to accepted Mediterranean practice. Pierre Gros (2003), however, has recently proposed that both archaeological and written evidence suggest that the wood in which the temple of Apollo on the Palatine Hill was situated was no mere precinct, but an integral part of the sacredness of the sanctuary as late as the reign of Augustus. In any case, Lucan can hardly be considered a reliable source for such a late occurrence of woodland animistic worship. His translator in the Loeb edition describes his primary intent as being to make his readers' 'flesh creep' (Duff 1928, xiii). In addition, he appears to feel that the grove had as much significance to the Massaliotes as to the Gauls. The following passage describes reactions to the destruction of the sanctuary by Caesar's troops at the siege of Massalia:

The peoples of Gaul groaned at the sight; but the besieged men [i.e. the Greeks] rejoiced; for who could have supposed that the injury to the gods would go unpunished? (Pharsalia III.445-50).

Lucan's vagueness as regards of the role of the grove as a ritual focus in the area around Marseilles suggests that such sites may have been redundant by the first century AD.

Human sacrifice too had only just become obsolete in Rome in the early first century BC, (Piggott 1968, 99), and the practice of judicial execution, such as the ritualised public strangulation of Vercengetorix, Caesar's defeated enemy in the Gallic Wars, while staged as an execution, had many overtones of the human sacrifice of war-victims (Ian Armit, pers comm). The vehement castigation of the practice in Celtic societies may be a hostile reaction to recognising a commonality with a barbarous people who are supposedly benefiting from Rome's civilising influence. Similarly, as regards headhunting, which even the softest primitivists found difficult to reconcile with their Noble Savages, and about which Diodorus wrote: to continue hostility against the dead is bestial (5.29), the Romans may be protesting too much. There is little written evidence of any similar practice in Greek or Roman texts, apart from an obscure reference in Livy at the end of Book 5 which includes a speech by Camillus encouraging the Romans to rebuild Rome after 390 (5.54.7):

Here is the Capitol, where a human head was once found, and it was predicted that there should be the head of the world and the fount of empire.

Whatever the significance of this remark, there is clear iconographic evidence of Roman head-taking in war depicted on, for example, Trajan's Column (fig 7.15), and it is perhaps this which informs the simplistic interpretation of the Celtic practice as the mutilation of their enemies and the acquisition of heads as war-trophies.

4.13 Misunderstandings of Native Practices?

Headhunting and Ancestor Cults

The classical texts make a clear link between warfare and headhunting, a link which, as discussed in Chapter seven, is borne out by many anthropological examples (e.g. Maxwell 1996; Gibson 1990; Whitehead 1990). The idea of head-taking as a symbol of courage, is imperfectly demonstrated by the following example from Silius Italicus, as Quirinius had already been attacked and seriously wounded by another man in a situation which palpably fails to reflect Vosegus' bravery:

Vosegus cut off his [Quirinius's] head from behind, and carried off the helmet hanging by its plume with the dead man's head inside it, and hailed his gods with the war-cry of his nation" (Punica IV, 215-220).

Nevertheless, all the sources interpret this practice simply as the taking of war-trophies with no consideration of any more spiritual meaning (Ritchie and Ritchie 1995, 54). Modern day headhunting practices on the other hand suggest any number of more complex ritual functions (ancestor worship, fertility etc.). Furthermore the archaeological evidence suggests that the Classical authors have misinterpreted, or at least grossly simplified the custom. As discussed more fully in chapter seven, more than one form of treatment and display is demonstrated in the remains, with some skulls having undergone remodelling and decoration. The sculpture, by its highly formalised nature, also appears to characterise an elaborate grammar involving ritualised headhunting, with representations of armed warriors in stylised poses holding sets of severed heads. A reference in Silius to the death of Sarmens, one of Hannibal's Gaulish allies, may contain an oblique allusion to some such belief system:

Teutalus, pierced in the groin, fell before him, and the earth shook under his huge weight; and Sarmens next who vowed, if victorious, to offer to Mars his yellow locks - the hair that rivalled gold - and the ruddy topknot on the crown of his head. But his vow was unheard, and the Fates drew him down to the shades below with his locks unshorn" (Punica IV, 200-205).

In a footnote to the Loeb translation, J.D. Duff, explains that this passage probably refers to 'The knot in which the Gauls tied up their long hair [which] is often mentioned by Latin writers" (1927, 184). It also brings to mind the lock of hair by which some of the sculpted heads are grasped in the southern French warrior statues.

Arcelin interprets the apparently honorific preservation of some skulls as the desire to keep some important ancestors within the community for protection or as some other such ritual focus. This, he concludes, may be the Celtic custom alluded to above by Nicander, of spending the night near the tombs of their famous men in the belief that they will be granted oracles (Tertullian De Anima 57.10) (Arcelin et. al. 1992).

4.14 Suppression of Native Practices?

Both written and archaeological evidence suggest that the deliberate suppression of indigenous religious practice in southern France. Strabo explicitly states that human sacrifice and the displaying of enemy heads was an obsolete practice in Gaul in his time (4.4.5):

The Romans put a stop both to these customs and to the ones connected with sacrifice and divination.

The archaeology is, in many ways, more eloquent than the stated policy. The unnecessarily violent destruction of ritual structures at Entremont and La Cloche, along with the defacement of the statuary and the scattering of sculptural elements around the sites, is strongly suggestive of an iconoclastic event. This may have simply been intended to demoralise the enemy, however it may also indicate that the Romans identified a link between these ritual observances and political hierarchy among the Saluvian federations.

After the establishment of the Roman Province, warrior sanctuaries and other similar ritual spaces seem to disappear from native settlements in the lower Rhône Valley. The larger domestic structures in the region adopt Roman architectural and decorative features, while smaller households were dispersed throughout the countryside much as they had been at the end of the Bronze Age. Both textual references to the suppression of sporadic native revolts, (e.g. Jullian *Historia Philippicae* III, 120-21), and the remains of the desolation of indigenous political and/or ritual centres and the establishment of permanent Roman garrisons, are indicative of a final and rather ruthless subjugation of indigenous power. As mentioned above, Ebel suggests that, by the time of the Roman Civil War, Massalia was exceptional in its exclusion from Roman cultural colonisation of Transalpine Gaul (1976, 101).

The kind of religious hybridism, linking native and Roman religious deities in iconography and sanctuaries which is found in some of the Western Territories does not seem to develop in southern France. This phenomenon has traditionally been interpreted as Romano-Celtic amalgamation (Green 1989, 73), and recently postulated as a covert indigenous resistance which attempted to subsume the colonising gods (Webster 1997). Neither standpoint appears apposite to the apparently total disappearance of the political and religious identities of southern Gaul.

4.15 Summary

The history and ethnography of the peoples of southern France as recorded by their Greek or Roman neighbours vary as to their reliability and relevance. The ethnographies, as laid out above, suffer much from repetition, convention and the misunderstanding of indigenous practices witnessed by Greek and Latin authors. Nevertheless, the creation of the 'Celtic Warrior' of popular imagination has its origins in these accounts. From the earliest encounters with the populations of southern France in the sixth century, to Caesar's Gallic wars half a millennium later, the Celts were primarily adversaries in war. It is little wonder that, as Freeman asserts, the Mediterranean world saw the Celts as 'warriors par excellence' (2002, 1). Some of the customs described by Poseidonius, in particular headhunting, miss much of the subtlety of the practice, but essentially recognise the relationship between the display of heads and warfare, and the importance of this practice in the social and ritual life of southern Gaul. The longevity of this particular Celtic anecdote, which appears, not only in Poseidonius, but in a plethora of historians and poets, indicates its importance; however, a fuller and more nuanced picture of its function in Celtic, or at least Saluvian, warfare must be left to an analysis of the archaeological data.

Even as an approximate historic outline, these texts only consider the events which involved Mediterranean participation, with just intermittent and perhaps dubious references to embedded warfare amongst the native populations. Moreover, gaps in the records, especially before 390 BC, mean that even major events involving Massalia may be missing completely. The Mediterranean version of events may indicate periods when warfare in southern France is highly organised and externally directed, such as during the Roman intervention in the final decades of the second century BC, however, as regards small-scale, inter-tribal, indigenous warfare the historic record of the Mediterranean is essentially deficient. Native conflict, in particular small-scale and early wars are nowhere alluded to apart from oblique references to the populations of Massalia being in conflict with the warring Celtic and Ligurian tribes (Justinus 43.3.13; Strabo 4.1.5).

Nevertheless, while the details and the exact dates of even the most trustworthy of the histories are questionable, taken together the accounts can provide a framework of events which can be used as a 'working model' against which to consider the archaeological evidence for warfare. Moreover, despite its short-comings the documentary evidence also gives some insight into the nature of the Gaulish-Mediterranean civilised-tribal zone, which, as discussed in chapter two, has been shown to have potent repercussions on internally directed warfare in tribal societies (Haas 1990, 171-72; Ferguson 1984, 43-45). Even so, it is left to archaeology to distinguish these more internally-directed patterns of warfare: their scale and impact, how sustained they were, and how localised? and, potentially, the overarching societal correlates of war in southern France during this period. The following chapter begins this investigation at the core of social organisation, exploring possible evidence for war in the settlement record of the indigenous and colonising populations of southern France.

5.1 Introduction

This chapter examines a range of indigenous sites in the Lower Rhône Valley in order to assess the ways in which warfare might be reflected in the settlement record. Indications of warfare or violence at these settlements may be derived from direct evidence for aggression, which comprises demonstrable indications of violence and destruction, and indirect evidence for aggression, which includes factors such as location and defensive architecture. In addition to the conventional forms of evidence just mentioned, settlements will be examined for more notional indications for warfare, formulated from evidence for socialisation for mistrust, or the breakdown of supra-familial trust-relationships within communities, which anthropological study (Ember and Ember 1992) has suggested may be a strong correlate of recurrent warfare.

5.2 Background

As described briefly in chapter three, urbanisation and socio-political complexity initially occur among the indigenous populations of southern France between the late-seventh and early-first centuries BC (Arcelin et al. 1983; Marty 2002; Verdin 2002, 146-47; Walsh and Mocci 2003, 63-66).

The preceding period, around the end of the Bronze Age and beginning of the Iron Age (*c.* 900 – 725 BC), is, as yet, poorly understood. The latter part of the Bronze Age in Provence was characterised by a variety of different site types and sizes, from isolated upland farmsteads to small agglomerated villages of perhaps a hundred occupants on the plains (Garcia 2000, 49; 2004, 28). The greater reliance on pastoral farming (relative to the Second Iron Age) and use of perishable building materials, have suggested a society which had not yet adopted fully-permanent sedentary occupation (Garcia 2004, 27- 34; 2005, 170; Py 1993, 78-79).

The 'Proto-Urban' Period (c. Sixth and Fifth Centuries BC)

Changes in settlement organisation and architectural techniques began to occur in Mediterranean France at the end of the seventh century BC (Arcelin et al. 1983, 138-39; Bouloumié 1990, 33-36; Guichard and Rayssiguier 1993; Dietler 1997, 313-14) around the time when Etruscan and Greek mercantile activity intensified along the coast, and became more widespread around 600 BC with a permanent Greek presence in the region (Py 1993, 83-92; Gras 2004, 229-30). From this time, particularly along the coasts, and hills overlooking plains, some nucleated settlements were established within stone ramparts, and houses were constructed of permanent materials, on stone footings with adobe or mud-brick walls (Arcelin et al. 1983, 141; Duval 1999, 105-06).

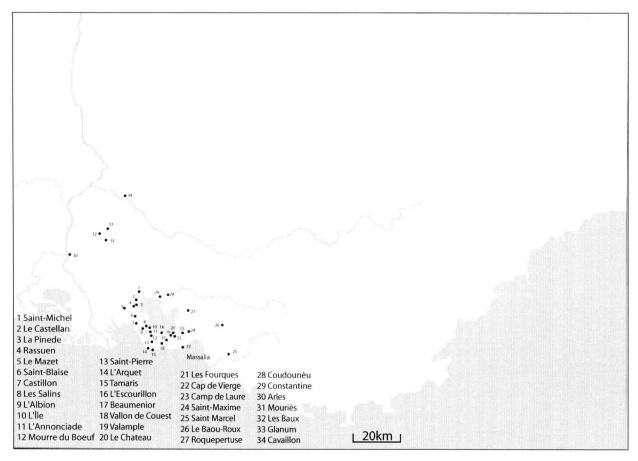

1 Saint-Michel	13 Saint-Pierre	
2 Le Castellan	14 L'Arquet	21 Les Fourques	28 Coudounèu
3 La Pinede	15 Tamaris	22 Cap de Vierge	29 Constantine
4 Rassuen	16 L'Escourillon	23 Camp de Laure	30 Arles
5 Le Mazet	17 Beaumenior	24 Saint-Maxime	31 Mouriès
6 Saint-Blaise	18 Vallon de Couest	25 Saint Marcel	32 Les Baux
7 Castillon	19 Valample	26 Le Baou-Roux	33 Glanum
8 Les Salins	20 Le Chateau	27 Roquepertuse	34 Cavaillon
9 L'Albion		
10 L'Île		
11 L'Annonciade		
12 Mourre du Boeuf		

Fig 5.1: Principal excavated settlements of the sixth and fifth centuries BC

Domestic space in these new agglomerations was arranged in blocks or 'islets' comprising quadrangular, one-room houses, either sharing a common axial wall or closely abutting one another (Duval 2000, 124-25). These islets were constructed in single rows of houses (often against the inside of the rampart), or in double rows of houses arranged back-to-back, and laid out in a grid-like network of streets (Dietler 1997, 310-11). Settlement sizes remained variable for around the next two centuries. Many new foundations were larger than any previously known; the ramparts at Saint-Blaise for example enclosed over 5ha, however, most, like L'Arquet and Tamaris, fell between 0.5 and 1.5ha. In Provence this move towards permanent nucleated settlement developed most quickly in the vicinity of Massalia, most notably in the areas south and west of the Étang de Berre in the commune of Martigues (Chausserie-Laprée 2005; Garcia 2000, 52). These sites occupied a variety of environments, mostly concentrated on the coast, on the fertile plain of Saint-Julien south of the Étang, or on the agricultural lands bordering the proliferation of salt-water lagoons (the so-called 'zone of lagoons') west of the Étang de Berre (Trément 2000; Chausserie-Laprée 2005) (fig 5.1).

While these changes in settlement organisation were widespread, and occurred in a variety of ecological environments, they generally coincided with both economic and commercial factors. Garcia (2000, 51-55) has suggested that (in the absence of direct environmental evidence) more productive farming techniques and the beginnings of supra-subsistence agriculture can be inferred from the appearance of pithoi (grain storage jars), and the introduction of iron tools.

These new settlements were generally established on coasts, rivers and other axes of communication, and they proliferated in the areas of permanent Greek colonisation (*ibid.*); although the relationship between the early Phocaean trade and indigenous socio-economic organisation is still not entirely understood (Py 1993, 107-111; Gras 2004, 229-30). The transformations in Gaulish settlement patterns during this period have been seen as the emergence of 'proto-urbanisation' (Chausserie-Laprée 2005, 58; Garcia 2005a, 169) in Provence, as the centralisation of power implied in the construction of these settlements and the adoption of permanent architecture, are considered to lay the foundation for later urban development.

This is not to suggest that these sites necessarily functioned as centres of administration, production, redistribution or any of the possible criteria listed by Collis (1976, 3) to classify a small or budding town. Indeed, the function, and, to some extent, the classification, of these early examples of urban or proto-urban settlement in southern France, have not yet been determined. Models have been constructed regarding the social, economic and political function of various 'central places' (Cunliffe 1995, 91) in Iron Age Europe. The central European 'princely' seats of the sixth and fifth centuries BC, such as the Heuneburg or Mont Lassois, for example, appear to have acted primarily as centres of

production, with rows of small workshops for the manufacture of prestige goods within substantial ramparts (Kristiansen 1998, 255-63). Around a century later, Danebury hillfort in Hampshire, while producing evidence for a variety of activities, seems to have had large areas given over to grain storage (Cunliffe 1995, 98-101).

Cunliffe points out, however, that the function of apparently prominent sites, and the social role of those occupying them, can only be partially derived from internal characteristics, such as house sizes and artefactual evidence (*ibid.*, 89-90). Interpretations of how the Hallstatt centres, Wessex hillforts and other such places may have functioned in the political and economic lives of the Iron Age societies that created them, have been developed, based on the way in which these centres interacted with other sites in the surrounding landscape. Settlement hierarchies and burial traditions in particular, potentially hold much information regarding social structure. Models which seek to explain the social, economic and political systems represented by the Hallstatt centres, for example, are subject to some disagreement (fig 5.2 and fig 5.3); but such hypotheses are only possible because similar sites with associated burials have been located and explored throughout the region.

At Danebury (Cunliffe 1995), the site's involvement in the control and redistribution of agricultural surpluses, trade goods and raw materials, was identified through evidence for production (large-scale grain storage and craftworking areas) and exchange (traded material, iron currency-bars and weights). The full complexity of this role, as well as the trajectory of the political life of the centre, and the ease (or otherwise) with which it held this power, however, became discernible only when the hillfort was viewed within the context of the occupation of the surrounding landscape. Reference to contemporary settlement hierarchies and land-use, highlighted the distinctions between Danebury and other settlements with which it may have relationships of clientship and/or competition (fig 5.4).

Studies such as those just mentioned (Cunliffe 1995; Kristiansen 1998) regarding the manner in which prominent centres functioned within their wider socio-economic landscape, make use of data sets not currently available in the archaeological record of the Lower Rhône Valley. For example, the highly stratified society which controlled the Hallstatt centres was to a great extent inferred from the very differentiated burial wealth discovered in the barrows which surrounded them (Wells 1984, 22). There are, however, no corresponding data in Provence, where human remains are rarely found, and no formal cemeteries have, as yet, been discovered (see chapter seven). The evidence is, again, elusive as regards settlement hierarchies, and the differences in the use and size of settlements, for a number of reasons. Provence has not been subject to wide-ranging prospection and exploration, and archaeological investigations have tended to concentrate on individual, enclosed sites, as

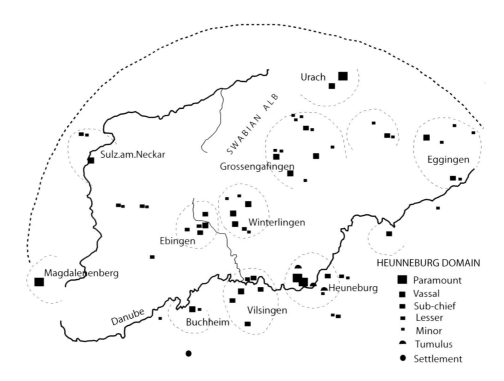

Fig 5.2 Frankenstein and Rowlands (1978) suggest a model of paramount chiefs governing vassal chiefs, all utilising one princely centre (Kristiansen 1998, 264).

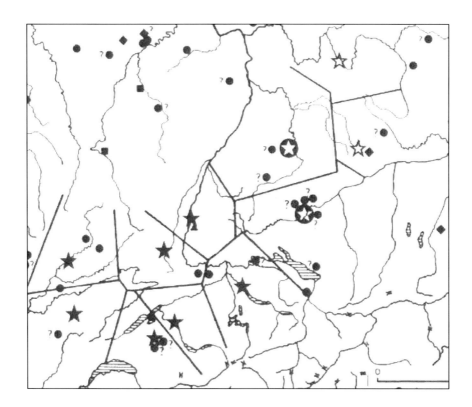

Fig 5.3 According to Härke's (1979) interpretation, each princely residence (indicated by a star) is an autonomous polity (Kristiansen 1998, 265) surrounded by a network of smaller vassal chiefs (ibid. 266).

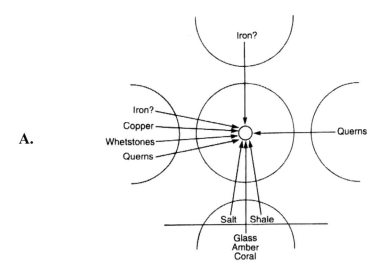

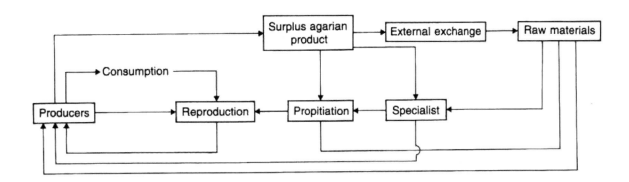

Fig 5.4 Model demonstrating how Danebury may have functioned as a centre for trade, production and administration.

A. shows the exchange items and raw materials coming into the site from north, east, west and (below the line) over seas.

B. demonstrates the cycle of consumption, production and redistribution through which exchange goods and agrarian surpluses may have been controlled from the centre.

(From Cunliffe 1995, 94 and 99)

these are more visible than other forms of settlement. Other than the enclosed settlements, occupation of the region is almost undetectable at present. Low-density or semi-sedentary occupation, on coastal sites and in caves, is occasionally recognised from pottery scatters and hearths (Dietler 1997, 310; Chausserie-Laprée 2005, 44-45), however, in the Mediterranean climate, such fragile remains are susceptible to erosion (Kevin Walsh 2005 pers. comm.). This has discouraged attempts at prospection so that, due to the poor survival of the remains, and the belief that the recovery of such limited remains is possible or worthwhile, settlement hierarchies in the region are poorly represented.

The result of this is that, firstly, attempts to reconstruct the role of seemingly central places within the lives of the population as regards socio-economic organisation, politics and warfare, are limited. It is not possible to ascertain how these centres compared and interacted with the more ephemeral, contemporary occupation sites. Secondly, it is impossible to compile a comprehensive or fully representative settlement distribution map as a large section of the population must remain invisible. This may be particularly problematic for the fourth and third centuries, when existing nucleated settlements decline or fall into disuse, and new foundations are rarely more than 0.5ha (two or three times smaller than settlements of the previous two centuries).

The function of a settlement, both in terms of its use and its relationship with the surrounding territory, also has some bearing on its classification. The term 'oppidum' is frequently used in relation to the enclosed, nucleated Iron Age settlements of the Midi. Sometimes (e.g. Py 1990; Dietler 1997) the term is used specifically for settlements of the fourth century BC onwards, which, as described below, demonstrate a more consistent settlement organisation. More often, however, the word 'oppidum' refers to the broader tradition of enclosed, nucleated indigenous settlement which began around 600 BC (e.g. Guichard and Rayssiguier 1993, 231-56; Chausserie-Laprée 2005, 67; Garcia 2005b). The oppida of southern France were much smaller and lacked the range and scale of craft production evident on the oppida of temperate Europe and southeast Britain which emerged in the final centuries BC (e.g. Buchsenschutz and Krausz 2001, 292-94; Collis 1975, 87, 97, 115). Nevertheless, the processes which led to the conception of both these traditions were in many ways similar. In both cases power centres were founded on the control of regional or local resources and developed increasingly complex political organisation through their participation in external trade networks (e.g. Cunliffe 1976; Collis 1976, 19; Haselgrove 1976). In the context of this study, the term 'oppidum' is used to indicate an indigenous settlement of the sixth to first centuries BC, enclosed by ramparts and comprising the kind of densely occupied surface of regularly laid out islets divided by a grid of (usually very narrow) streets, described above. The term does not necessarily indicate a particular size of settlement, an upland location, involvement in supra-subsistence trade or production, or even a defensive nature (although, as will be seen, the

locations chosen for these settlements suggest that defence was a consideration in all periods).

The Consolidation of Urban Settlement (c. Fourth and Third Centuries BC)

The Second Iron Age (*c.* 475/50 – *c.* 50 BC) saw further changes in settlement patterns and organisation in the region (fig 5.5). These included a more intensive and centralised form of occupation by the indigenous populations. Throughout the fourth and third centuries BC in particular, there was a dramatic re-ordering of indigenous settlement throughout Mediterranean France. During these centuries in Provence, settlement types became less varied. Small dispersed farmsteads all but disappeared (Garcia 2004, 82), and the enclosed hilltop agglomerations, while being the most commonly recognised form of indigenous settlement, declined in both size (usually 0.5ha or less) and numbers (e.g. Gazenbeek 2000, 86; Marty 2002, 151) (fig 5.5). Many of the early indigenous sites in Martigues for example, were either abandoned or greatly reduced in size, while new enclosed settlements were founded in the uplands on the outskirts of Marseilles (Chausserie-Laprée 2005, 46; Dietler 1997, 317). The apparent decrease of people residing in permanent settlement may indicate a return to more mobile and, therefore, less visible, modes of living. However, the more common interpretation of this phenomenon is a demographic 'recession' (Arcelin 1992, 320; Py 1993, 153-54; Garcia and Bernard 1995, 116).

Conversely, during the same two centuries, there appear to be significant innovations in settlement layout and function, and an increasing rationalisation of space. Houses became much more uniformly rectilinear, and islets and streets were arranged to a more regular and overarching schema while settlements were more densely occupied (Garcia 2004, 84-101). In addition, there appears to be more evidence for food processing and craft-working areas within oppida at this time. Oil and wine-presses, for example, became widespread (e.g. Chausserie-Laprée 1990, 61), and some settlements in the Massaliote hinterland produced evidence for communal storehouses (e.g. Gantès, L.-F. 1990, 80). These innovations were concomitant with a movement (what Garcia terms a 'rural exodus') of indigenous settlement towards the large Greek commercial sites such as Massalia, Agde or Emporion (Garcia 2000, 55 and 2004, 82-89).

The fourth and third centuries BC are considered to be a period of 'urban concentration' (Garcia 2004, 84). The formalisation of occupied space and the inclusion of elements of storage and production indicate a more centralised political leadership which had some control over the distribution and exchange of the community's surplus. Despite their small size, these fortified settlements in the Massaliote hinterland appear to have achieved a supra-subsistence economy, creating the potential for specialisms and involvement in commercial networks. It is this function which Garcia (2000, 55) calls

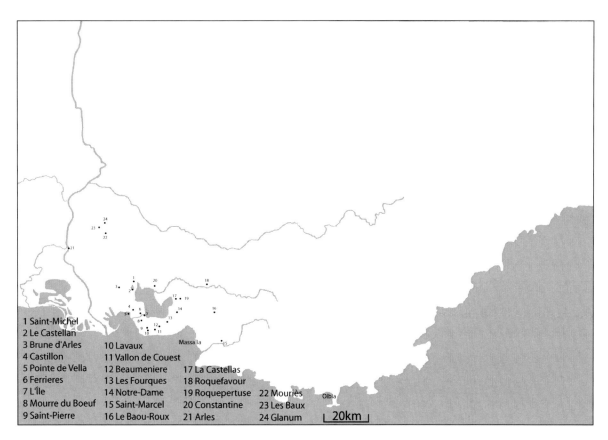

1 Saint-Michel
2 Le Castellan
3 Brune d'Arles
4 Castillon
5 Pointe de Vella
6 Ferrieres
7 L'Île
8 Mourre du Boeuf
9 Saint-Pierre
10 Lavaux
11 Vallon de Couest
12 Beaumeniere
13 Les Fourques
14 Notre-Dame
15 Saint-Marcel
16 Le Baou-Roux
17 La Castellas
18 Roquefavour
19 Roquepertuse
20 Constantine
21 Arles
22 Mouriès
23 Les Baux
24 Glanum

Fig 5.5: Principal excavated settlements of the fourth and third centuries BC.

the 'criteria for...and origin of' urbanisation: involvement in economic activities over and above subsistence.

Indigenous Centres of Power and Complexity (c. Second and First Centuries BC)

The early-second century BC marked yet another transformation of indigenous modes of settlement (fig 5.6). This period began with the sudden abandonment of many of the oppida described above, often following a violent assault. The second and first centuries BC saw the emergence of large agglomerations of up to 6ha (Dietler 1997, 317). Some early sites, like Saint-Pierre (Lagrand 1979), were expanded and/or restructured, while others, like Pierredon (Boissinot 1990) were constructed on sites with no evidence of prior permanent settlement structures. Along with an increase in occupied space, these new oppida exhibited an unprecedented monumentality, with large and elaborate fortifications and formally laid out public and ritual areas with statuary (Arcelin et al. 1992). An intensification in craft-working activities appears to have taken place on many of these sites, and the remains of mechanised oil-presses, and metal-working and pottery workshops have been recovered on several sites from this period (e.g. Brun 1993, 101-05; Hummler 1993, 130; Dufraigne 1999, 69). Domestic architecture also became more complex, with the addition of multiple rooms, usually within the traditional schema of abutting rooms, but sometimes (e.g. at Lattes in Languedoc) in the form of units arranged around an inner courtyard (Dietler *ibid.*; Thierry Janin

2004 pers. comm.). The fortified agglomerations which emerged after 200 BC were the largest and most complex manifestations of the indigenous settlement tradition since the beginning of the Iron Age. This tradition came to an end following the Roman intervention around the last quarter of the second century BC, after which time most indigenous settlement in the Lower Rhône Valley became either heavily Romanised (e.g. Glanum) or reverted to the dispersed patterns of the Late Bronze Age (Dietler 1997, 319; Chausserie-Laprée 2005, 54). One of the last examples of this kind of fortified, nucleated settlement in the region was the oppidum of La Cloche, which was established after the Roman annexation of the Provence, and destroyed and abandoned around 50 BC (Chabot 1983 and 2004).

These centres were the largest and most complex form of indigenous settlement in the southern French Iron Age. Some areas of production, in particular the processing of olives and grapes, appear to have been carried out on an almost industrial scale in the final centuries BC (although many households still continued to carry out grinding cereals or wood and leather-working, as domestic activities) (Arcelin 1993, 84). It should be mentioned that the centres of Mediterranean France were still much smaller than many of the better known oppida of Central France, southern Germany and adjacent areas (Collis 1975, 33 and 1984, 149-50). Nevertheless, the southern French oppida of the second and first centuries BC appear to have been urbanised, political, economic and administrative centres, where indigenous elites controlled

41

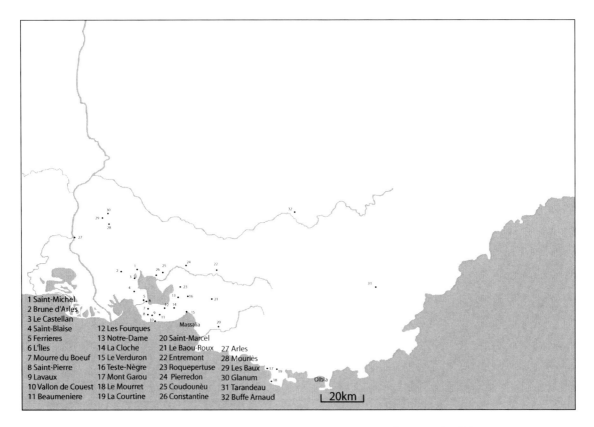

Fig 5.6 Principal excavated settlements of the second and first centuries BC.

agricultural and other resource production for trade with the Greek colonies, indigenous communities and, increasingly throughout the second century, with Rome (Chausserie-Laprée 2005, 53).

5.3 Methodology

Ten indigenous settlements have been selected for detailed analysis from the three main periods of development in the southern French indigenous settlement tradition outlined above. As discussed, a comprehensive understanding of the southern French settlement record has been limited by past excavation strategies and problems of differential survival. Until very recently, archaeological study has focused almost exclusively on enclosed settlements. The few attempts at more wide-ranging investigation (e.g. Walsh and Mocci 2003) have been hampered by environmental conditions. The effects of the arid climate and violent storms of the Mediterranean have left early occupation surfaces either eroded (Kevin Walsh pers. comm.) or masked by colluvial or alluvial deposits (Gazenbeek 2000, 85). In addition, the restructuring which took place on some of these sites in the second and first centuries BC often greatly disrupted earlier levels. Some sites, such as Saint-Pierre, Saint-Blaise and Glanum also saw subsequent Hellenistic or Gallo-Roman occupation which, besides being destructive of earlier phases, became the main focus of nineteenth- and early twentieth-century excavation campaigns. The result of this stratigraphic damage and a former preference for Greek and Roman archaeology, is that the range of indigenous activities

carried out between the sixth and the third centuries BC is not always well accounted.

In the analysis which follows, the sites are listed chronologically, taking the main period of enclosed, nucleated settlement as the 'start date' for sites with a long history of occupation (but referring to previous unenclosed occupation of the site where relevant). Settlements have been selected for well-defined chronologies, and relatively good levels of information on house types, architectural floor-plans and settlement layout. It should be mentioned that settlements where the entire surface area has been excavated are rare; and that chronologies are often based on relative sequences rather than absolute dating, so that, dates for phases of occupation, abandonment and episodes of aggression are approximate.

Factors Considered in Analysis

Direct evidence for aggression

This comprises the evidence commonly interpreted as representing an episode of siege-assault at these sites, and may include one or more of the following criteria:

- level of destruction (demolished buildings, objects broken in situ (statues, storage jars etc.), finds of scattered projectiles)
- fire damage to ramparts
- widespread areas of burning
- evidence of violent death

Indirect evidence for aggression

The intention (as opposed to the potential) to defend each site is quantified on a scale of 'small' to 'very large' on a combination of three main criteria:

- width of ramparts
- strategic positioning of fortifications (comprising such factors as the positioning of ramparts, outworks etc., outlined in full in Appendix 2)
- natural defensive location

Socialisation for mistrust

The cross-cultural study by Ember and Ember (1992), described in chapter two recognised a strong correlation between embedded warfare, and fear and mistrust within the community in non-state level societies. In the ethnographic samples on which the Embers' study was based, 'socialisation for trust' was a statistical variable, coded by Herbert Barry as 'Mutual confidence... [referring to] confidence in social relationships, especially towards community members outside the family' (Barry et al. 1976, 114). This psychological state was seen in such things as children being free to enter any house in a village, or leaving possessions unguarded. Mistrust was classified as evidence for the antitheses of these qualities and included the securing of personal belongings and an increase in superstition and witchcraft accusations (Carol Ember pers. comm.). Socialisation for mistrust could be seen in, for instance, the fragmentation of 'families' into small nuclear units rather than communal or extended-family groups. Thus, fear and mistrust among members of the same community (as a correlate of recurrent conflict within the wider landscape), might be seen in small, carefully-delineated, family living-spaces within a settlement, as much as in fortifications enclosing the entire community.

An attempt will be made to operationalise these findings by the examination of the following:

- the measurement of household size to identify fragmentation of the communities into 'nuclear' family groupings
- access analysis to gauge the relative freedom of movement between households and around the settlements
- changing levels of communality within the sites (presence or absence of communal areas)
- superstition within the sites (the presence of cultic features at the personal or domestic level)
- intention to secure personal goods (presence of keys, locks, small caskets, and evidence of personal goods being hidden or secured e.g. caches of coins)

Identifying Fragmentation through Household Size

Defining 'Nuclear' Households

This study uses the definition of 'nuclear' household as a household comprising two adults and their children (Laslett 1974; Blanton 1994, 5; Byrd 2000, 66). Nevertheless, there are still variations in the numbers which might constitute a nuclear as opposed to an extended family. In a recent review of the ethno-archaeological methodology applied to these questions, Byrd (2000) has suggested that attempts to determine household organisation often contradict one another because they are rarely based on extensive and empirical ethnographic data (e.g. Bar-Yosef and Meadow 1995; Hole 1987, 82; Voigt 1990, 3). According to Byrd (2000, 64-65), one exception to this is Flannery (1972) whose conclusions were based directly on a wide-ranging study of modern African pastoralists. Flannery (1972, 42) defined nuclear households as usually comprising a family of two adults and one or two children (i.e. three or four people in total).

Measuring Household Size

The approximate size of a household is generally calculated from a dwelling's floor-space. One of the most-quoted formulae for determining household size assigns $10m^2$ of floor-space for each person (Naroll 1962, 587-89). This figure has been borne out in independent applications to societies from a variety of different environmental circumstances by LeBlanc (1971, 210-11). Naroll's methods have been criticised because his allocation of a standard amount of floor-space per individual applies an over-simplified formula to large households, which may have been employed in a wider variety of domestic activities (Norbeck 1971). In an extended household, residents might be expected to undertake additional tasks, thus, each additional occupant would increase the range of craft or domestic activities in which a household could engage. Consequently an extended household might require larger areas for tools and food storage. Naroll's formula was subsequently refined by Sherburne Cook (1972, 16) who allocated $7.62m^2$ for each of the first six individuals and an additional $30.48m^2$ for each additional occupant. The application of both figures may be used to give a likely range for the size of a nuclear household. Thus, three individuals at Cook's formula ($3 \times 7.62m^2$) and four individuals at Naroll's formula ($4 \times 10m^2$) gives a range of $22.86 - 40m^2$ for estimating the floor-space required to accommodate a nuclear household. The results of this analysis are not intended to yield a precise determination of the numbers of residents or the classification of those households, as even the most complete settlement evidence is limited by too many ambiguous factors. Nevertheless, the quantification of household sizes throughout the period may provide a general 'scale order' of fragmentation, and, therefore, potentially, mistrust within communities.

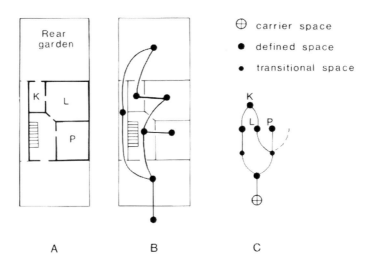

Fig 5.7

A.   Plan of the lower floor of a modern house.

B.   House-plan overlain with an unjustified access diagram.

C.   Justified version of the plan showing the levels of obstruction (doorways, lanes) which divide spaces.

(After Foster 1989, 41).

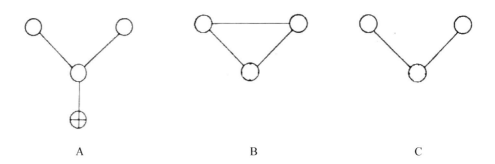

Fig 5.8 This analysis is mainly concerned with the extent to which domestic spaces are separated or connected.

A.   The Justified access diagrams used in this analysis take this form, showing the 'carrier space' i.e. outside the settlement gate, and consequently, the levels of movement into the settlement and between households.

B.   Spaces which are directly accessible produce a 'distributed' diagram.

C.   Those which are entirely unconnected produce a 'non-distributed' diagram.

(After Foster 1989, 43).

Identifying Communality through Freedom of Movement between Households

Access Analysis

One means of investigating levels of spatial integration is through access analysis, for example along the lines explored by Foster (1989), particularly as regards her examination of obstructions to movement within settlements (fig 5.7 and fig 5.8). Foster's intention was to use access analysis as an indicator of social structures, but her methods may also indicate aspects of socialisation for mistrust.

5.4 The Settlements

The following section describes ten sites in the Lower Rhône Valley, noting direct and indirect evidence for aggression (where present) and highlighting potential indications for socialisation for mistrust.

5.5 Tamaris (*c.* 600 BC - *c.* 550 BC)

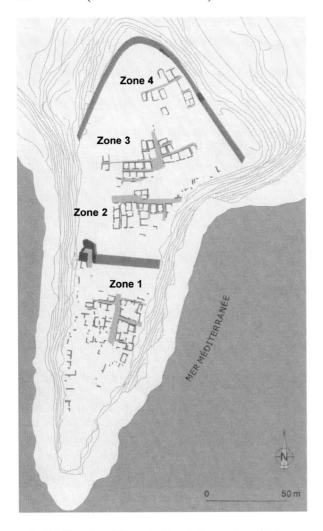

Fig 5.9 Site plan of Tamaris. The darker areas on both ramparts show the fully-excavated sections including the area of the curved baffle-gate which has been thoroughly investigated. Access analysis was carried out on Zones 2 and 3, which have the most complete plans. (After Chausserie-Laprée 2005, 90).

Site Description

Tamaris was an *éperon barré*, or promontory fort stretching around 300m into the Mediterranean just south of the Étang de Berre. Two parallel ramparts cut off a roughly triangular area of around 1.5ha comprising four 'zones' of occupation. All of these elements were laid out and constructed contemporaneously around 600 BC (Chausserie-Laprée 1990, 55; *ibid.* 2005, 89; Duval 2005).

History of Research

The northern part of this site has undergone an extensive and recent campaign of excavations (1998-2004) by Sandrine Duval; however, due to chronic natural erosion and an earlier and a more limited campaign of work directed by Charles Lagrand in 1960-61, the southern tip of the promontory (zones 0 and 1) is only partially understood (Chausserie-Laprée 2005, 90; Duval 2005, 134).

Direct Evidence for Aggression

There is evidence for the destruction and burning of some structures around 550 BC. Nine Greek bronze arrowheads with bent points were recovered from the area of the gate of the inner rampart. No indigenous weapons or projectiles were found and no human remains were discovered. The site was reoccupied after this event, but abandoned shortly afterwards (Chausserie-Laprée 2005, 68).

Indirect Evidence for Aggression

Natural Defensive Location

Tamaris was afforded protection on two of its three sides by cliffs, rising between 11 and 20m above the sea (Chausserie-Laprée 2005, 90) (fig 5.9). The site was exposed to the Mistral and to an almost equally icy and destructive eastern wind, as well as to erosion on its seaward sides (*ibid.*, 62).

Northern Rampart

The northern rampart was originally around 2.2 - 2.8m thick, although at least part of the wall had been increased by a further 1m, creating either a staircase or some reinforcement to the inner side of the fortification (Chausserie-Laprée 2005, 66-67).

Southern Rampart

The southern rampart, protecting the remaining half hectare of the settlement at the tip of the headland was around 6m in width, and was protected by a sophisticated system of defence, centred on a complex gateway flanked to the east and west by two large towers between 5.5 – 6m² (fig 5.10). Around the middle of the sixth century the entrance was blocked off transforming the gate into the southern part of the settlement into a curved dead-end (fig 5.11) (*ibid.*; Duval 1999, 106).

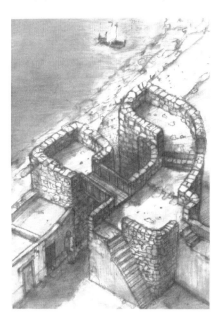

Fig 5.10 Reconstruction drawing of the curved baffle-gate in the southern rampart at Tamaris. (Drawn by Denis Delpalillo, from Chausserie-Laprée 2005, 67).

Fig 5.11 The curved baffle-gate at Tamaris from the south, showing blocking of the gate with rubble and un-worked stone. The rough blocking appears hurriedly constructed compared with the fine, dry stone-construction of the rest of the rampart (Photo: the author).

A. Entrance into the site.

B. Blocking stones barring the entrance-way.

C. The entrance road into the runs along the edge of the cliff for a few meters and slopes down into the cove below.

Strategic Positioning of Fortifications

The north rampart ran along the highest point of the promontory, and would have controlled the only possible approach to the site from dry land. The western tower of the southern rampart was constructed at the edge of the cliff and covers the only cove around the promontory which is accessible from the sea. The eastern tower on the other side of the gate was extended northward and bends across the access path, creating a curved baffle-gate.

Socialisation for Mistrust

Household Size

Analysis of house size and access at Tamaris are limited to zones 2 and 3, as these are the most completely recorded areas of the settlement. House sizes vary from $12.5m^2 - 31.2m^2$, however, the average house size is $21.5m^2$, below the size range ($22.86 - 40m^2$) suggested for small or nuclear households.

Access Analysis

There is no evidence for intercommunication between units (fig 5.12 and fig 5.13) in the areas examined, although this may not be universally applicable. Duval noted that none of the zones of settlement at Tamaris utilised the same layout (2000, 124), Lagrand, for example recorded that zone 1 comprised blocks of units arranged around central courtyards, however, current site plans are incomplete (*ibid.*, 2005, 137).

Ease of movement around the settlement is limited to circuits of streets around one or two housing units. In general, movement between households involves crossing several transitional spaces (access routes, streets). The division of the settlement by the southern rampart would further increase this separation of domestic units if zone 1 was included in the analysis. In addition the excavator noted that at Tamaris, as at Saint-Pierre and L'Arquet, some streets had been blocked off with low stone walls, transforming them into impasses, however, only at Saint-Pierre is this specifically described.

Communality

No formal public or communal spaces have been incorporated within the settlement. Street widths are variable, with wide central axes of circulation but streets between house rows which are barely wide enough to walk along.

Superstition/Securing of Personal Goods

A communal cultic area for the deposition of small metallic objects was established in the dead end created by the blocking of the baffle gate in the middle rampart (Chausserie-Laprée 2005, 66-67).

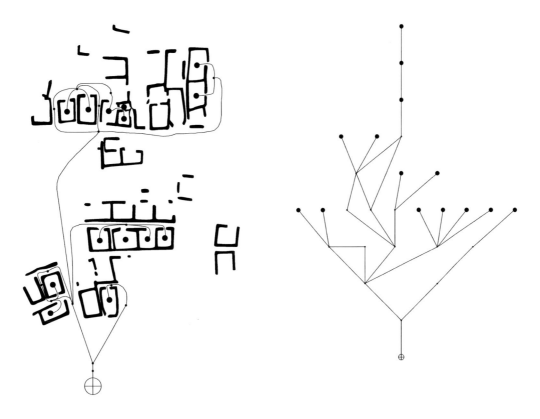

Fig 5.12 (Left) Tamaris zones 2 and 3: unjustified access diagram.

Fig 5.13 (Right) Tamaris zones 2 and 3: justified access diagram.

5.6 Saint-Pierre (*c.* 550 BC – *c.* AD 100)

Site Description

Saint-Pierre was established on a small hill in the fertile plain of Saint-Julien, 6km from the Étang de Berre and the Caronte channel to the north, and Cap Couronne and the Golfe de Fos to the south. The settlement has undergone continuous occupation from its foundation in the mid-sixth century BC to the present day, where it is still the site of a church and small village. In its original construction in the sixth century BC, the rampart enclosed around 1.5ha, however, in the fourth century the settlement was enlarged, and a new fortification took in a further 2000m². Saint Pierre continued to expand outwith the walls until the first century BC when the Roman annexation signalled the end of indigenous settlement-forms in the region (Chausserie-Laprée 1990, 58; 2005, 16-17).

History of Research

Continuous and intensive occupation of the hill has done much to destroy or disturb the Iron Age remains at Saint-Pierre, and greatly limited the excavations carried out in the 1960s by Charles Lagrand (1979; Chausserie-Laprée

1998, 92-93). Nevertheless, recent rescue excavations, instigated by Jean Chausserie-Laprée between 1998 and 2001 (1998; 1999; 2000; 2005), have explored all of the accessible space on the hill, and produced a good general sequence of the settlements development (fig 5.14).

Direct Evidence for Aggression

Saint-Pierre has produced no direct evidence for assault.

Indirect Evidence for Aggression

Natural Defensive Location

Saint-Pierre is the only prominence in the middle of the hollow plain between Saint-Julien and the sea. The hill is small, and its chief natural defensive value may lie in its panoramic views which cover the major routes of communication between Marseille, the southern coast and the Étang de Berre.

Early Rampart

To the east the rampart varied in width, from 1m to 2.2m thick, with two quadrangular towers, 7m x 5.2m, and 7m² (this last was periodically enlarged to 12m x 9.40m) (Chausserie-Laprée 1998, 93; 2005, 80).

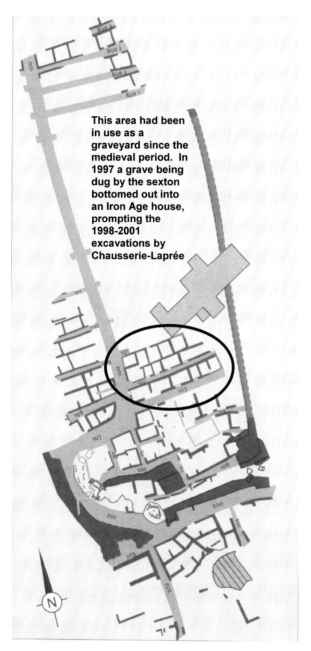

This area had been in use as a graveyard since the medieval period. In 1997 a grave being dug by the sexton bottomed out into an Iron Age house, prompting the 1998-2001 excavations by Chausserie-Laprée

Fig 5.14 Site plan of Saint-Pierre showing the Iron Age settlement, Roman construction and subsequent medieval and modern structures. The access analysis was carried out on the most complete part of the settlement plan (circled) and the entrance way. (After Chausserie-Laprée 2005, 92).

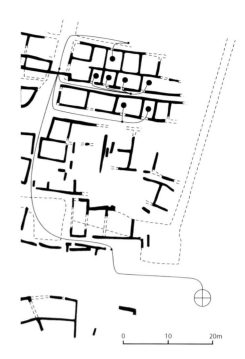

Fig 5.15 Saint-Pierre: unjustified access diagram from the southern part of the settlement.

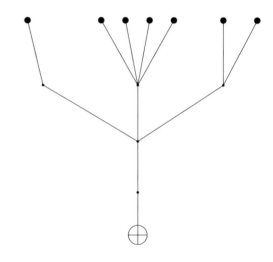

Fig 5.16 Saint-Pierre: justified access diagram from the southern part of the settlement.

Southern Rampart

Around the beginning of the fourth century BC the original rampart was destroyed to extend the settlement and a new fortification, 0.7m - 1.6m wide, was built to the south. Three towers from this new rampart survive: two bastions projecting 1.2m and 4m from the rampart wall and a round tower of 30m² (Chausserie-Laprée 89-84).

Strategic Positioning of Fortifications

The eastern bastion projected from the rampart next to the main gate, restricting vehicles on the approach-road just prior to entering the settlement and the round tower overlooks the road to Marseille (Chausserie-Laprée 2005, 82).

Socialisation for Mistrust

Fragmentation

Houses in the initial phase of settlement had an average area of 20m^2. At the end of the fourth century when the settlement was expanded, houses inside the ramparts were partitioned, reducing each surface area to an average of around 8m2 (Chausserie-Laprée 2005, 130-31).

Access Analysis

Saint Pierre produces a completely non-distributed diagram, and the uniformity of the known settlement layout suggests that this lack of communication between households was replicated throughout the settlement (fig 5.15 and fig 5.16). In the second century BC in the northeast area of the site some streets were blocked by low walls up to 60cm high, creating blind alleyways in the vicinity of the eastern rampart, one of which was associated with the ritual deposits described below (Chausserie-Laprée 2005, 109).

Communality

A structure just over 30m^2 was uncovered dating to the foundation of the settlement. The building has been interpreted as a hypostyle or symposium hall and included a double-doored entrance, an interior portico and an installation in one corner with several hearths. It is not clear if this room continues in use throughout the life of the settlement (Chausserie-Laprée 2005, 92).

The only other indication of public foci are the ritualistic elements added to the ramparts and the street barrier described below, however, these are quite small spaces and are not formally delimited. The original ramparts utilised ancient stelae in the construction of both wall and towers, which may have constituted a public ritual focus, although no actual space is delimited (Bessac and Chausserie-Laprée 1992; Chausserie-Laprée 1999, 101).

Towards the end of the first millennium BC, five blocks or columns of white, dressed sandstone, with cephaliform niches were ranged along the exterior face of the rampart at the eastern part of the curtain wall. A pillar bearing a carved human head was also found (possibly nearby, though no context was recorded). In the heart of the settlement, a small wall which blocked the end of one street in the northeast of the village became the depository for around twenty sherds of imported ceramics, each inscribed with the Gaulish name *RITUMOS*. This has been described as a temporary 'sanctification' of the oppidum to Ritumos, a god or local hero (Chausserie-Laprée 2005, 109, 239).

Superstition/Securing of Personal Goods

Individual or domestic ritual practices are in evidence in all parts of the settlement throughout its occupation.

These comprise deposits, particularly in house floors; notably the remains of young animals, found in their hundreds, as well as small votive objects. Numerous indications of household libation rituals, again buried or half-sunk into household floors, were also discovered. Personal amulets made from fossils, corral and shell, have been found in a large proportion of the houses. One sixth or fifth century house also produced a Bronze Age stele in its foundations (Bessac and Chausserie-Laprée 1992; Chausserie-Laprée 2000, 123; Chausserie-Laprée 2005).

5.7 L'Arquet (>450 BC – <300 BC)

Site Description

L'Arquet was an *éperon barré*, established on a small promontory extending 75m into the Mediterranean two kilometres west of Tamaris. A rampart cuts the northern side of the site off from the mainland, enclosing the entire 0.5ha of the oblong headland. The settlement originally consisted of three rows of houses (two single, one double), however, much of this area has been destroyed by ancient quarries and natural erosion (fig 5.17) (Chausserie-Laprée 1990, 56; 2005, 28).

Fig 5.17 L'Arquet: much of the headland has collapsed into the sea; however, traces of the early, open-cast quarrying are still apparent. (From Chausserie-Laprée 2005, 28).

History of Research

The chronology of the settlement is known only through early excavations carried out during Charles Lagrand's investigation of a number of sites in the commune of Martigues between 1955 and 1962, and recently revised in relation to recent work in the region by Jean Chausserie-Laprée. The site is very badly damaged so that details about the ramparts and phases of settlement are more than usually incomplete. The urban layout at L'Arquet (and presumably the enclosure of the site) cannot be dated more firmly than between the middle and the end of the fifth century BC. For this reason the occupation of the site is designated the arbitrary date: >450 BC (Chausserie-Laprée 2005, 96-98).

Direct Evidence for Aggression

At some point between 450 BC and the end of the fifth century the settlement was abandoned following an episode of destruction and burning. The site was reoccupied but abandoned again at an unspecified date before the beginning of the third century BC, hence the designation: <300 BC (*ibid.*; 1990, 56).

Indirect Evidence for Aggression

Natural Defensive Location

The settlement of L'Arquet was protected on three sides by sheer, 12m high cliffs and exposing the occupants to a constant battery of violent winds and rough seas (Chausserie-Laprée 2005, 66).

Rampart

The remains of a defensive work run for 30m between the east and west cliffs and completely cut the settlement off from the mainland. Test pits through the several metres of collapsed rubble suggest that the entire defensive system was around 3m thick (*ibid.*).

Additional Indirect Evidence for Aggression

L'Arquet has produced some unusual indications for what might be considered aggressive intervention. Petrographic evidence reveals that the headland was being exploited as a stone quarry by the Phocaeans from the fifth century BC to supplement ongoing domestic and monumental building projects in Massalia. While the site was still in permanent occupation by an indigenous community, houses were cut through by sections of open-cast mining. Homes and parts of streets were progressively blocked off and abandoned as the quarrying advanced into the settlement. Subsequent to the definitive abandonment of the settlement in around 300 BC, L'Arquet was sporadically frequented by the indigenous population for fishing, but continued to be intensively utilised as a stone quarry by the Greeks (Chausserie-Laprée 1990, 56-57; 2005, 98).

Socialisation for Mistrust

Fragmentation

The first permanent settlement at L'Arquet comprised between 35 and 40 houses. Some surviving units in the middle row suggest almost square houses with an average surface area of 16m^2 (Chausserie-Laprée 1990, 56).

Access Analysis

The natural and deliberate destruction of so much of the settlement make L'Arquet a poor subject for access analysis. It has been observed, however, that like Saint-Pierre, some of the streets have been blocked by low walls, restricting movement around the settlement (Chausserie-Laprée 2005, 109).

Communality

There is no evidence for public or communal spaces in the settlement.

Superstition/Securing of Personal Goods

L'Arquet is described as having produced similar manifestations of household ritual to those at Saint-Pierre, including deposits of animals in house floors, however, published details of these are currently unavailable (Chausserie-Laprée 2005, 232).

5.8 L'Île des Martigues (*c.* 430 BC – *c.* 200 BC)

Site Description

L'Île was a fortified, artificial island constructed on reclaimed marshland at the point where the Caronte Channel (which runs from the Mediterranean) empties into the Étang de Berre (Chausserie-Laprée 1990, 59). The Caronte Channel was the foremost waterway between the Mediterranean and the heartland of indigenous settlement, around the zone of lagoons, and (presumably) around the shores of the Étang. The settlement began as a small agglomeration around 0.5ha in size in the fifth century but grew to more than three times that in the second century, when it continued to expand until its abandonment around 200 BC (Chausserie-Laprée 2005, 95-100).

History of Research

In the course of a sustained campaign of excavations carried out by Jean Chausserie-Laprée from 1978 to 2003 (Chausserie-Laprée et al. 1984; 1987; 1990; 1992; Chausserie-Laprée 2005, 16-24). L'Île has produced one of the most precise chronologies of any settlement in Iron Age Europe. Throughout its protohistoric occupation, the island continued to sink and the inhabitants were continually forced to build up the ground-level (Chausserie-Laprée 1990, 60-61), so that we now have abandoned layers every 50 or so years comprising house floors and street surfaces. These waterlogged layers have also resulted in the preservation of many perishable materials, such as painted wall friezes or stored foodstuffs, offering a unique window into the everyday lives of the indigenous populations of southern France (e.g. Chausserie-Laprée 2005, 161-84; Damotte 2004).

Direct Evidence for Aggression

L'Île has produced direct evidence for three episodes of destruction during its occupation. The first, in around 430 BC, while the settlement was still under construction, a second in around 360 BC and a final episode in around 200 BC which resulted in the ultimate abandonment of the site. In each case every house in the settlement was demolished and the site was subjected to a widespread burning. More than 30 javelin heads of Greek manufacture were discovered scattered around the site in the 360 BC destruction level (Chausserie-Laprée 2005, 87, 97-100, 129).

Indirect Evidence for Aggression

Natural Defensive Location

Its unstable situation created inconveniences for the occupants of L'Île, and defence was apparently a consideration in the choice of its location as seen initially in the utilisation of the lagoon in the construction of the defences (see below) and later in the extension of the site towards the east, further into the lagoon, rather than onto terra firma.

Rampart

The rampart varied in width from 1.45 to 2.6m and three large semi-circular towers, approximately 7m deep by 10m wide, were built against the western flank and appear to have been reinforced at different times throughout the first phase of settlement (Chausserie-Laprée et al. 1984, 27, 39-45; Chausserie-Laprée 2005, 85, 92).

Strategic Positioning of Fortifications

The main gate was located in the eastern wall overlooking the lagoon, and the one known postern faced the Caronte Channel to the south and was blocked off at the end of the fifth century, less than a generation after the site's foundation (Chausserie-Laprée 2005, 92).

Socialisation for Mistrust

Fragmentation

From its foundation the average household at L'Île measured 13m^2. In the rebuilding of the settlement following the *c*. 360 BC assault, the average floor-plan rose to 17.3m^2. Another 60-70 years of continued expansion saw a contraction of living space by the partitioning of contemporary houses so that by around 300 BC the average house was 10.3m^2.

These partitions were removed, however, around 50 years later, restoring the average floor space to its original size

of 13m^2 (Chausserie-Laprée et al. 1987, 43-48; Chausserie-Laprée 2005, 130-31).

Access Analysis

The unparalleled levels of preservation at L'Île present the opportunity to gauge changes in relative ease of movement around the settlement through several phases (Phase one: fig 5.18 and fig 5.19; phase two: fig 5.20 and fig 5.21; and phase three: fig 5.22 and fig 5.22). These phases, their chronology and where they lie in relation to periods of destruction and rebuilding, are outlined with the access analysis diagrams below.

The increasing restrictions of access around the site come from, apparently, incidental changes, such as the removal or blocking of secondary, transverse streets.

Communality

The first phase of settlement at L'Île (prior to 360 BC) has produced two open-air zones. One, a square 45m^2 in the northwest corner of the settlement; the other a gap in the middle of islet B, around 135m^2, which was gradually built-up in the course of the second century BC. Neither were formally laid out public spaces and seemingly served only as dumping grounds for domestic waste (Chausserie-Laprée 2005, 116-118).

Periods 3b and 4 encompass two sequential phases of reconstruction. Period 3b is the phase of rebuilding which took place after the 360 BC destruction. Period 4 equates to the phase subsequent to the Period 3b reconstruction and prior to the ultimate destruction of the site in 200 BC. The site plans for these phases are identical

After the 360 BC destruction and prior to the 200 BC destruction the site plan does not change significantly. Both periods produce a completely non-distributed diagram. These phases produce similar diagrams which are not entirely non-distributed, suggesting that the settlement plan did not deviate substantially after the 430 BC destruction.

Superstition/Securing of Personal Goods

Numerous small household ritual objects have been found. Several dozen animal deposits were uncovered in house floors in all periods of the settlement and evidence of libation practices was also common. In addition L'Île produced deposits of perforated metal objects and pieces of glass similar to ritual deposits found in other sites throughout the Midi, one unique adaptation being a murex shell still pierced by a large carpentry nail, which is dated to the beginning of the fourth century BC (Bessac and Chausserie-Laprée 1992; Chausserie-Laprée 2005, 200).

Several caches of coins have been discovered in all parts of the settlement and dating to all periods of occupation (Chausserie-Laprée 2005, 220).

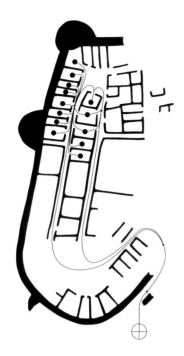

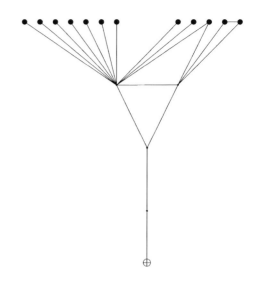

Fig 5.18 (Left) L'Île 'Period 2c'. This period and those immediately preceding it ('period 2a and 'Period 2b') encompass several phases of rebuilding but the site plans are substantially the same. Period 2a equates to the initial foundation of the settlement interrupted by the 430 BC destruction. Period 2b is the phase of construction which completed the foundation of the settlement. Period 2c (above left) signifies a period of widespread rebuilding (re-surfacing of streets and house-floors) just prior to 360 BC.

Fig 5.19 (Right) Justified access diagram of these phases.

These phases produce similar diagrams which are not entirely non-distributed, suggesting that the settlement plan did not deviate substantially after the 430 BC destruction.

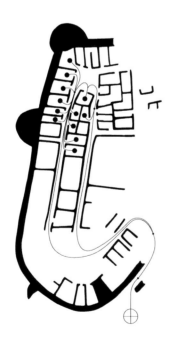

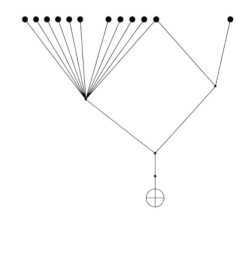

Fig 5.20 (Left) L'Île 'Period 3a' represents the phase of rebuilding and repair after the 360 BC destruction.
Fig 5.21 (Right) L'Île justified access diagram 'Period 3a', after the 360 BC destruction. The diagram indicates a slight increased restriction of movement around the settlement.

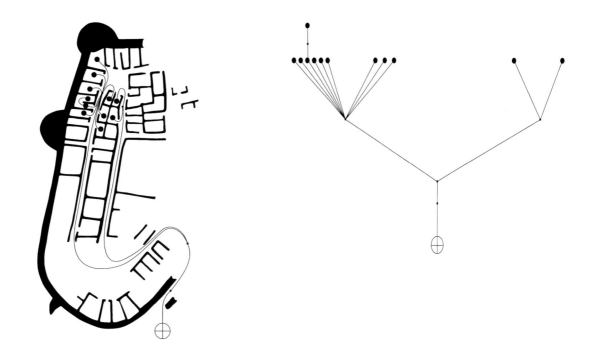

Fig 5.22 (Left) L'Île 'Periods 3b' and '4'. Fig 5.23 (Right) L'Île justified access diagram 'Periods 3b and 4'. (All plans taken from Chausserie-Laprée et Nin 1990, 38-39, published results of 20 years of investigation (1978-98) into the domestic arrangements at L'Île).

5.9 Notre-Dame-du-Pitié (*c.* 325 BC – *c.* 260 BC)

Site Description

Notre-Dame-du-Pitié sits on an upland plateau overlooking the Étang de Berre around 16km northwest of Marseille. The ramparts follow the natural morphology of the plateau running around 100m along each of its three sides and enclosing a roughly triangular area of around 3,750m^2 (Gantès 1990a, 73) (fig 5.24).

History of Research

Notre-Dame had been investigated briefly in the mid-1930s by Fernand Benoît, but more recent excavation has been carried out by Lucien-François Gantès in the late 1970s and 1980s. Restructuring of the site in antiquity and the middle ages, and modern road construction have all done much to damage the early archaeological levels. Households in particular are preserved only around the southeast rampart. However, Notre-Dame is still one of the best surviving examples of settlement from this phase, with major periods of occupation between 325 and 260 BC (*ibid.*).

Direct Evidence for Aggression

Notre-Dame has produced no direct evidence for assault.

Indirect Evidence for Aggression

Natural Defensive Location

The southern edge of the site ends abruptly in a sheer drop of almost 90m, while the northwest and northeast sides fall away in a gradual, easily accessible slope (Gantès 1990, 73).

Ramparts

An early phase of enclosed settlement (*c.* 400 BC – *c.* 325 BC) no longer survives, but a short length of wall 1.3m wide and the remains of a tower of undetermined dimensions beneath a later housing unit are still discernible. The main period of consolidation of the indigenous 'urban' tradition on the site (*c.* 325 – *c.* 200 BC) saw all three sides of the settlement surrounded by a rampart of varying thickness, from 2m at the northern end to 5.5m towards the southwest (*ibid.*,73-74).

Southern Fortifications

The southeast corner was protected by a sub-oval tower which projected 3.5m from the rampart wall. Later the southeast angle was further reinforced by a projection of undetermined dimensions and later again along the cliff, a bastion was added to the rampart which survives to 8m along the curtain wall.

Northern Entrance Gate

The main northern entrance gate was flanked to the west by a round tower 7m in diameter; and to the east by a square tower measuring 5m^2. In advance of this arrangement the northern point of the settlement was further protected by an outwork 1m wide (Gantès 1990a, 73-74).

Strategic Positioning of Fortifications

Some of the most substantial fortifications run along the southern cliff (suggesting they were constructed primarily for display), however, the more vulnerable northern gate was much more strategically protected. The positioning of the outwork, in particular, would have protected the gate against catapult shot (see Appendix 2).

Socialisation for Mistrust

Fragmentation

The first proto-urban settlement was established around 325 BC. Only one row of houses ('Islet 1') has survived from this phase, with single-room units around 11m².

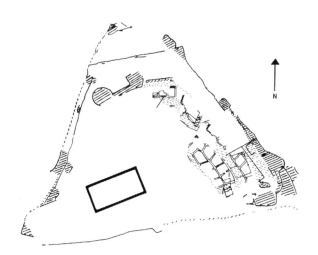

Fig 5.24 Site plan of Notre-Dame-du-Pitié.

Like many such sites in the region, subsequent use has destroyed much of the Iron Age settlement (the rectilinear structure is a medieval church).

Substantial ramparts run along the high southern edge of the site. Only islet 3 survives well enough for access analysis. (After Gantès 1990, 72).

Around 300 BC the old settlement was reordered. Islet 1 was extended and given new beaten-earth floors and clay plastering on the walls. The best-preserved housing block from this phase was 'Islet 2', a double range of domestic units along an axial wall, separate from Islet 1 by a wide street. For the most part this islet followed the same urban schema as before, and despite a rather haphazard layout, the households were almost uniformly 10 – 11m². One unit, however, was slightly larger than the others (14m²) and connected with its neighbour via a door in a partition wall (*ibid.*).

Access Analysis

The 300 BC 'Islet 2' at Notre-Dame produces a diagram with some distribution (occasioned by the double unit at the end of the islet) (fig 5.25 and fig 5.26).

Communality

The surviving streets at Notre-Dame were comparatively wide, between 2.4 to 3.8m and a third islet was separated from the other two by a small square of around 7m² (Gantès 1990a, 74).

Superstition/Securing of Personal Goods

The site produced no evidence of cultic practices, neither public religious monuments/deposits nor domestic ritual.

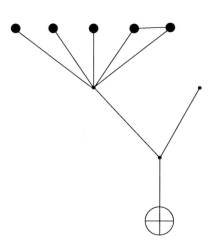

Fig 5.25 (Top) Notre-Dame-du-Pitié, islet 3: unjustified access diagram from the third/second century settlement. Fig 5.26 (Bottom) Notre-Dame-du-Pitié: justified access diagram from the third/second century settlement.

5.10 Roquepertuse (*c.* 250 BC – *c.* 200 BC)

Site Description

Roquepertuse occupies a plateau in the fertile Arc valley, around 15km from Aix-en-Provence and overlooking the Étang de Berre to the west. The site is best known as a ritual centre (discussed in more detail in chapter six). Elements of ritualistic lapidary including Bronze Age stelae and warrior statues from the fifth century BC (suggesting activity on the site for several centuries prior to the permanent settlement) were assembled into a 'hero sanctuary' in the third century, where human heads (real and simulacra), were displayed in a painted, stone portico. In the mid-third century BC the sanctuary was destroyed and the small domestic occupation accompanying it increased to its full extent of over 0.5ha (Boissinot and Gantès 2000).

History of Research

Isadore Gille of the Musée Borely in Marseille, commissioned the earliest investigations at Roquepertuse (carried out by his assistant, Henri de Gérin-Ricard) in 1919–24 and 1927. This work, which virtually ignored the settlement remains, was carried out primarily to retrieve elements of statuary for the museum, and the haphazard excavations and poor recording caused widespread damage to the stratigraphy of the site. The transfer of the Roquepertuse collections to the Centre de Vielle Charité inspired a renewed interest in the site (Lescure and Gantès 1991, 9-10; Gantès and Lescure 1992, 155). Consequently, between 1994 and 1999, Philippe Boissinot and Lucien-Francois Gantès conducted excavations which determined the extent of the enclosed area and defined several phases of occupation, destruction and abandonment from the Neolithic to the second century BC (Boissinot and Gantès 2000). Various other specialists have since analysed many of the ceramic, architectural and skeletal remains from Roquepertuse (e.g. Coignard and Coignard 1991; Delamare and Guineau 1991; Barbet 1991; Mahieu 1998) greatly adding to the understanding of the site.

Direct Evidence for Aggression

A little before 400 BC, a small Iron Age occupation came to an end with a sudden abandonment followed by widespread burning (Boissinot 1994, 164). The site was enclosed around the beginning of the third century BC and the 'sanctuary', accompanied by a small, nucleated settlement, was established; however, around 250 BC the statuary was destroyed (*ibid.* 2000). After much revision of the stratigraphy, this event is now considered to be independent of any destruction of the domestic settlement. Following this event, the settlement expanded to its full extent but contracted following a siege-assault around 225 BC evidenced by destruction, scatters of stone boulet and widespread burning (Boissinot 2004, 51). The subsequent, much smaller settlement was destroyed around 200 BC by incendiary. Some boulet have also been found in this level, but these may be residual finds (Boissinot 1998, 112).

Indirect Evidence for Aggression

Natural Defensive Location

Roquepertuse was established against the southwest flank of an imposing limestone eminence. The site slopes upwards from the southeast in a series of constructed terraces until it reaches the plateau which is bounded along its northern edge by a sheer 10m drop (Lescure and Gantès 1991, 10).

Ramparts

The initial (*c.* 300 BC) rampart survived only in the area around the entrance and consisted of a stone wall around 1.2m thick with the possible remains of a ditched area outside. A semi-circular tower, 4.8m in diameter partially survives to the east of the gate.

Not long after the initial construction, the rampart was reinforced and improved. The external, south side of the wall was increased by the addition of dressed stone facings, to around 2m in width (Boissinot and Gantès 2000, 252-56).

Strategic Positioning of Fortifications

The location would appear to be a perfect natural bolt-hole, with a single, sloping direction of approach which could be (and was) cut off with a stone rampart: easily accessible for those living and working in the countryside around the site, and defensible once inside. Nevertheless, the initial phase of fortification (which was associated with the sanctuary), appears to have been largely for display as much of the settlement was established beyond the limit of the rampart. Moreover, the gate was wide and easily entered via a wide staircase (Boissinot 1999, 118). The tower may also have been constructed primarily for display, as it was not supported by the rampart: a thin band of gravel (around 0.1m wide) separates the foundations of both (Boissinot and Gantès 2000, 255).

Socialisation for Mistrust

Fragmentation

The remains of two houses survive from around the second half of the fifth century, almost 150 years before the site was enclosed, both estimated to be around 24m^2 (Boissinot et Gantès 2000, 257). In the mid third century Roquepertuse reached its full extent. Three houses from this phase have almost identical surface areas (21.5m^2) (*ibid.*, 263) but these are the only well-defined domestic structures.

Access Analysis

Access routes and doorways from the settlement have not survived sufficiently to carry out access analysis.

Communality

The exact layout of the early third-century sanctuary remains a matter for consideration, however, its probable location, on a terrace at the top of a wide staircase, suggests that it was a public space. More prosaic communal spaces, such as workshops, are as yet unknown.

Superstition/Securing of Personal Goods

Small metallic objects (pierced metal plaques, beads etc.) found scattered throughout the third-century levels, are similar to those found in votive deposits in settlements throughout southern France. Evidence for domestic libation rituals and two infant 'foundation deposits' have been found in house-floors from the third century BC, however, it is unclear whether these date to the early third-century settlement associated with the sanctuary, or the later oppidum (Boissinot 2004, 52).

5.11 Teste-Nègre (*c.* 250 – *c.* 200 BC)

Site Description

Teste-Nègre was established on a small eminence 13km northwest of Marseille. The settlement overlooked the fertile plain of Saint-Victoret-Marignane, and the natural route between Marseille, the Golfe de Fos and Saint Blaise, which gave access to many the coastal and lagoon ports and constituted an important part of the road between Italy and Spain (Gantès 1990b, 79). Artefactual evidence suggests a limited occupation between 375 and 260 BC, however, no structural evidence predates the foundation of the oppidum in *c.* 225 BC. This settlement was destroyed and abandoned *c.* 200/190 BC (*ibid.*, 83).

History of Research

Early, small-scale excavations were carried out on the northern part of the site of Teste-Nègre in 1904-06 by Michel Doumens, and by L. Malzac in 1935, and between 1950 and 1955 Louis Chabot and Jean-Baptiste Féraud explored the length of the rampart. More recently, excavations on the occupied surface were carried out between 1975 and 1976 by Lucien-François Gantès. While much information regarding the site is still unpublished, the current state of research regarding the site has been synthesised by Gantès (1990).

Direct Evidence for Aggression

Rounded slingshot pebbles were found both in heaps and scattered all over the settlement. The remains of a much-corroded iron sword were also found in this level, directly below that of a wide-spread incendiary dated to around 200 BC. The site was subsequently abandoned (*ibid.*, 83).

Indirect Evidence for Aggression

Natural Defensive Location

The hill falls away abruptly on three sides while the southeast is joined to a plateau in the Nerthe hills, of which Teste-Nègre is an outlying foothill. The oppidum was established on the northwest flank of the hill, exposing the settlement to the full force of the Mistral.

Ramparts

The rampart was around 2m wide with one tower 2.8m by 5.7m, which was subsequently enlarged to around 2.8m by 10m along the length of the rampart.

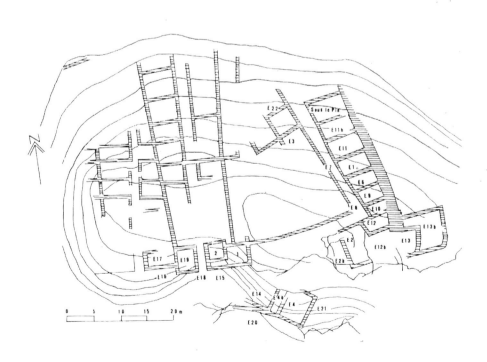

Fig 5.27 Site plan of Teste-Nègre: No formal plans other than this schematic drawing of the site have been published as yet. (From Gantès 1990, 78).

Strategic Positioning of Fortifications

Teste-Nègre was fortified along the only accessible approach to the settlement, creating an éperon barré oppidum. The sole tower was situated at the highest point of the site, and (although the entrance gate itself is unknown) appears to guard the probable narrow access route to the site (Gantès 1990b, 79-80).

Socialisation for Mistrust

Fragmentation

The settlement comprised around 50 houses, each of which contained at least one dolia for food or water storage, and the remains of wine amphorae. These were long and narrow, of regular size, 8m^2 on average, and separated by narrow streets of no more than 1m in width (Gantès 1990b, 78-81).

Access Analysis

Due to the lack of detailed published plans (showing doorways etc.) it has not been possible to carry out access analysis at Teste-Nègre.

Communality

Teste-Nègre produced no evidence of formally laid out public spaces, however, some of the buildings were given over to the storage of food and drink (*ibid.*).

Superstition/Securing of Personal Possessions

Teste-Nègre was not the site of a sanctuary and produced little evidence for a formal ritual space or furniture. The site produced much fine, imported ceramics and personal items of decorative metalwork (bracelets, rings etc.); however, finds of amulets, keys or caskets are not recorded in the 1990 synthesis or artefactual analyses (Gantès 1979a; 1979b).

5.12 Le Baou-Roux (*c.* 200 – *c.* 125 BC)

Site Description

La Baou-Roux occupied a flat plateau on one of the outlying hills of the Etoile Mountains halfway between Aix and Marseille above the fertile plain of Bouc-Simiane. The site had been occupied sporadically since the Neolithic, but the earliest permanent settlement dates from around 400 BC and was abandoned around 350 BC with no signs of violence. A subsequent settlement was established in the late-third century BC and destroyed around 200 BC, only to be replaced by a large and complex oppidum of around 4ha which was itself destroyed and definitively abandoned in 125 BC (Boissinot 1990, 91-93).

History of Research

Apart from an early and very poorly carried out exploration of the site in 1903 by G. Vasseur, Le Baou-Roux remained unexplored until the 1960s and 1970s

when excavations were instigated by the landowner. Like many of the larger sites in the Midi, La Baou-Roux has not been excavated in its entirety, and the early levels are greatly disturbed by the foundation of the oppidum. Nevertheless, its chronology, in particular the period of fortification and ultimate destruction, (*c.* 200 BC – *c.* 125 BC) has been well documented in more recent investigations carried out under Philippe Boissinot between 1983 and 1986 (*ibid.*).

Direct Evidence for Aggression

The level dated to around 200 BC produced evidence for the demolition of structures followed by widespread burning. A second destruction level dated to around 125 BC was marked by the demolition of structures, ceramics broken *in situ*, and widespread scatters of boulet (Boissinot 1990, 93). Two iron lance heads of differing types, a javelin point and a Greek bronze arrowhead have also been found around the site, as has a decorated bronze helmet which bears the marks of two heavy blows from an unidentified weapon; however, it is unclear whether these all date to *c.* 125 BC, or if some are residual finds from the *c.* 200 BC level. Neither is it clear whether these were used by the assailants or the defenders of the settlement (*ibid.*, 97-98).

Indirect Evidence for Aggression

Natural Defensive Location

La Baou-Roux joins the Etoile Hills from the south, and is only accessible from this direction via a small valley which cuts through the plateau. For the most part the hill is free-standing, bounded by an irregular series of cliffs around 300m high (fig 5.28).

Ramparts

The site was enclosed by ramparts around 2.5m wide at the beginning of the second century BC, at the period of its greatest expansion. These were doubled to 5m wide at the southern part of the plateau.

Strategic Positioning of Fortifications

The double-width rampart to the south barred the only entrance onto the plateau (Boissinot 1990, 91-93).

Socialisation for Mistrust

Fragmentation

The pre-200 BC settlement appears to have been limited in area and the major reordering of the site for the foundation of the enclosed settlement in the second century BC, has largely destroyed earlier levels. The average household in the phase of the oppidum was 25m^2, however, Boissinot states that many had roof terraces (1990, 93). The bases for this assertion is not stated, however, in other settlements, such as Entremont or La Cloche, a second storey or roof terrace is sometimes inferred from steps along an exterior wall of a house. If this is the case at La Baou-Roux, some household sizes would have been above, or at least close to the top of the size range suggested here as indicating a small or nuclear family.

Access Analysis

Le Baou-Roux produces a moderately distributed diagram (relative to other settlements in the region). Direct movement between households or rooms is not apparent, but connections between 'defined' spaces indicates an ease of movement, via 'transitional' spaces (circulation routes and transverse streets), around the settlement (fig 5.29 and fig 5.30).

Communality

While the pre-oppidum settlement (late-third century BC - *c*. 200 BC) survives only in part, there was evidence for communal areas for food storage and processing. The oppidum (*c*. 200 BC – *c*. 125 BC) produced no indication of any public building, space or monument.

Superstition/Securing of Personal Goods

La Baou-Roux produced no signs of a sanctuary or any other form of public foci, ritual or secular. Domestic religious or ritualistic objects, however, were frequent and, in some cases, unique. Some of these are ambiguous: a small stone cube, for example, described as a 'die' with zoomorphic carvings might be a gaming piece. A fine, wheel-thrown pitcher from Massalia which was found on the site was inscribed with a Gallo-Greek female name: *AEIOUITAI KONGENNOMAROS. M. M.* Bats who studied the piece sees it as a theonym and interprets this item as belonging to some domestic devotion (Boissinot 1990, 97).

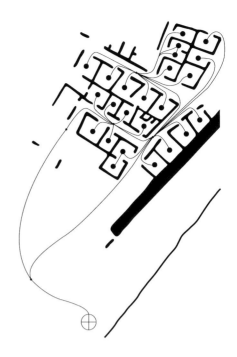

Fig 5.29 (Left) Le Baou-Roux: unjustified access diagram.

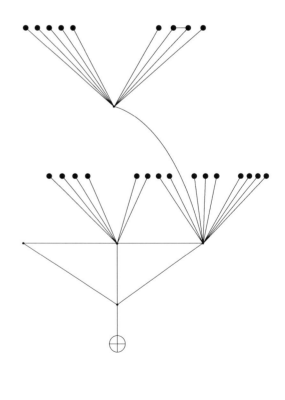

Fig 5.30 Le Baou-Roux: justified access diagram.

The diagram is relatively distributed, however, according to Boissinot (1990, 93) restrictions of access between houses and around the site increased near the end of the second century. These changes took place prior to the c. 125 BC destruction, when some of the streets were narrowed by around half, and others were completely blocked off.

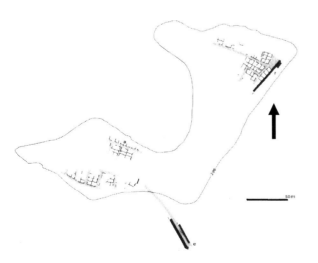

Fig 5.28 Le Baou-Roux. The scale indicates 50m.

The access analysis was carried out on the most extensively excavated area in the northeast. The hill can only be accessed by the southern pass. (After Boissinot 1990, 91).

5.13 Entremont (*c.* 175 – *c.* 90 BC)

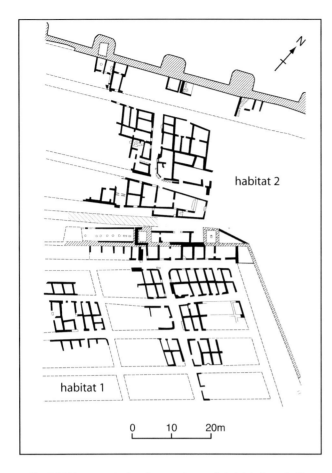

Fig 5.31 Entremont, showing most complete islets from both phases of enclosed settlement used for access analysis. (From McCartney 2006, 106 after Arcelin 1990, 102).

Top right-hand islet, habitat 1, Ville Haute (c. 175 – 150 BC)

Top right-hand islet, habitat 2, Ville Basse (c. 150 – 90 BC)

Site Description

The oppidum of Entremont was founded on a roughly triangular-shaped hilltop plateau, 3km from Aix and around 15km from Marseille. The hill overlooks the fertile Arc valley and the ancient 'Herculean Way' which crosses Provence from east to west (Arcelin et al. 1990, 101). The site underwent two successive phases of enclosed settlement. The first settlement, the Ville Haute, was established around 175 BC and comprised an area of approximately 9,000m^2. In around 150 BC, this early phase was replaced by a more complex oppidum, the Ville Basse, almost four times larger (Py 1993, 172). The size and complexity of the site, indicates that Entremont was an important centre in the second century. Its location (close to Aix) and the chronology and sequence of its episodes of warfare, correspond closely with the historical texts regarding the Roman intervention.

Consequently Entremont is believed to have been the capital of the nascent Saluvian federations of the second century BC (Arcelin et al. 1990).

History of Research

Antiquarian investigations had been carried out at Entremont since the early nineteenth century, mostly with the intention of recovering lapidary fragments. Many pieces of sculpture were dug up by German forces occupying the hilltop during the Second World War, prompting the first formal excavations between 1946 and 1967 under the supervision of Fernand Benoît, although these are considered flawed in terms of methodology and interpretation (e.g. Wheeler 1958, 211-212). In 1972, François Salviat carried out investigations in areas already explored by Benoît, however, the site is now best known through the meticulous and long-term work carried out by Patrice Arcelin, which began in 1984 and continues to the present (Arcelin 1993, 57; André and Charrière 1998).

Direct Evidence for Aggression

The Ville Basse was destroyed around 125 BC. Smashed and defaced elements of statuary and Roman boulet found scattered around the entrance road are believed to date to this level. The settlement was reoccupied, but abandoned after a second destruction, dated to around 90 BC, and comprising the destruction of parts of the Ville Basse, and scatters of boulet and pila inside the main entrance to settlement (Arcelin et al. 1990, 105; Dufraigne 1995, 137-38; 1999, 67-69).

Clay sling stones, believed by Benoît (1975, 231) to be indigenous, were also found in the Ville Basse, in stashes, ready for use and in a workshop where they had been manufactured.

Indirect Evidence for Aggression

Natural Defensive Location

The plateau is bounded by two sheer cliffs to the south and a sloping hillside to the north, where the main fortifications of both settlement phases were constructed.

Ramparts

The first rampart was 1.5m wide with the remains of at least three regular quadrangular towers around 6m by 6.5m. The second rampart was 3m wide with ovoid bastions, around 9m by 6m, along the curtain wall (Arcelin 1993, 68).

Strategic Positioning of Fortifications

Both ramparts barred the approachable north and northeast sides of the settlement.

Socialisation for Mistrust

Fragmentation

Households in the *c.* 175 BC settlement are very uniform with an average floor-size of 10m^2. House-sizes in the Ville Basse varied greatly, with an average range between 15 – 34m^2 (Arcelin et al. 1990, 104). Some houses had an upper storey (Arcelin 1993, 72) (as opposed to the kind of simple roof-terrace found at some sites such as L'Île). If the average surface-area of the second settlement at Entremont is doubled to take account of the extrapolated extra floor-space, the average domestic space comes to between 30 and 68m^2 (Arcelin et al. 1990, 104), suggesting that many houses in the later settlement comprised internal floor-space above the range given here for small or nucleated households (McCartney 2006; McCartney 2008).

Access Analysis

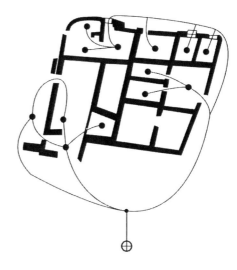

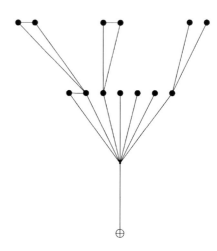

Fig 5.34 (Top) Entremont Ville Basse: unjustified access diagram (c. 150 BC). Fig 5.35 (Bottom) Entremont Ville Basse: justified access diagram (c. 150 BC). The diagram is deeper (perhaps suggesting more social complexity) and much more distributed. Arcelin et al. (1990, 104) note up to five interconnecting rooms in some islets.

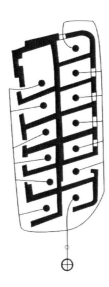

Communality

The streets of the Ville Basse are around 2-3m wide (two-three times as wide as the first oppidum) (Arcelin 1993, 72). Communal spaces, such as workshops and areas for food storage and processing are entirely absent in the first settlement, but occur frequently in the second (e.g. Brun 1993; Arcelin 1993, 97).

The chronology of areas associated with the display of iconography and the observance of communal rituals is still a matter for debate and will be discussed further in the following chapter (many of the elements pre-date both phases of occupation). Most of the material associated with areas of communal ritual, (such as the hypostyle hall described in the following chapters) was

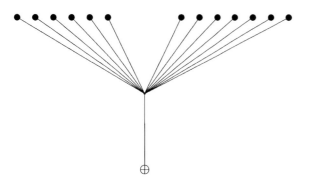

Fig 5.32 (Top) Entremont Ville Haute: unjustified access diagram (c. 175 BC). Fig 5.33 (Bottom) Entremont Ville Haute: justified access diagram (c. 175 BC). The diagram for the early phase of settlement is entirely non-distributed.

found in an area which would have been situated outside the ramparts of the first settlement and along the interior entrance road of the second oppidum (Arcelin et al. 1992).

Superstition/Securing of Personal Goods

A ritual deposition of small, pierced metal and glass items was found in a pit within one of the domestic units in the Ville Haute. No personal amulets or remains of domestic ritual were apparently associated with domestic spaces in the later settlement.

Pits hollowed out from the angles of houses for the concealment of coins and small valuables have been found from both phases of settlement. Some had been robbed, however, others still contained small stashes of drachmas and Massaliote obols. Numerous keys have also been retrieved; although is not clear whether these belong primarily to one particular phase of settlement (Willaume 1993).

5.14 La Cloche (*c.* 100 – *c.* 50 BC)

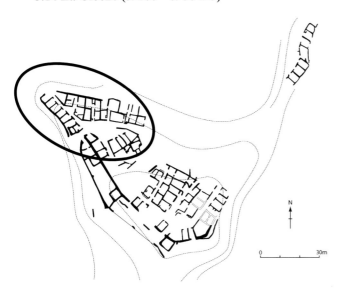

Fig 5.36 Simplified plan of La Cloche showing area around the entrance used for access analysis (below). (After Chabot 2004).

Site Description

The oppidum of La Cloche was situated on a hilltop 240m high, overlooking the long-established route between Massalia and the Étang de Berre. An initial phase of enclosed settlement, established around 125 BC took in about 1500m². In 110-100 BC the oppidum was enlarged by around five times its original size. The surface area of the existing settlement was almost completely re-used as material for terraces and steps constructed to counteract the incline of the hilltop (Chabot 2004, 143). This succeeding settlement was

destroyed around 50 BC. While the site is indigenous in both form and ritual practices (Marty 1999, 209), numerous Phocaean trade items, which might have been more easily obtained from the Gaulish hinterland, have led to suggestions of close links with Massalia, however, the exact relationship is unclear (Marty 1999, 210; Chabot 2004, 33) (Fig 5.36).

History of Research

La Cloche has been the object of an exhaustive campaign of excavation from 1974 to 2000, recently published by Louis Chabot (2004).

Direct Evidence for Aggression

In the mid-first century BC all the buildings were wrecked and filled with debris, and Roman projectiles were strewn in the streets of the oppidum along with the belongings of its inhabitants (Chabot 1997). Human remains were found at La Cloche, including a near-complete skeleton, various cranial fragments and stray bone found at different points around the site (Chabot 1994; 1983, 51), many of which appear to have been associated with the Roman assault (discussed further in Chapter 7). A large number of weapons were discovered scattered around the site, particularly around the entrance-way. These included two sword fragments, a dented and pierced buckler, 14 lance-heads, 11 pilum points, 10 lead and iron projectiles and 715 stone boulet. Several stockpiles of natural, stone pebbles were also recovered. One projectile, a lead catapult shot, came from the second-century BC layer, the rest came from the *c.* 50 BC destruction level (Chabot and Feugère 1993).

Indirect Evidence for Aggression

Natural Defensive Location

The settlement, which occupied the northern flank of the hilltop, was crescent-shaped, with both 'horns' pointing northward and down-slope. This crescent is set into a hollow on the hills summit, and defended by a sheer drop on the north side and a rocky crest to the south (Chabot 1990, 119; 2004, 59).

Ramparts

The earlier, *c.* 125 BC rampart was around 1.25m thick. One irregularly built tower, around 4m by 4.6m, also survived by being incorporated into a larger tower from the second phase of settlement. The *c.* 100 BC rampart was 1.2 to 1.3m thick with three square towers of around 5m by 5m (Chabot 2004, 62-64).

Strategic Positioning of Fortifications

The second phase of fortifications was equipped with a supporting wall around 1m thick and two of the later towers each have an additional wall running at right angles between the rampart and the supporting wall, thus cutting off the road encircling the oppidum and restricting the approach to one main direction. In addition to the main entrance, the rampart was broken by three posterns, one of which had been hurriedly blocked around the time of the *c.* 50 BC destruction (*ibid.*)

Socialisation for Mistrust

Fragmentation

The two surviving house-floors from the *c.* 125 BC settlement are 11.50m^2 and 13.50m^2. The average surface area for the second phase of settlement is 16m^2. In the summit area some houses had two storeys, calculated as having between 35m^2 and 45m^2 of living space (Chabot 2004, 69).

Access Analysis

The relative depth of access at La Cloche indicates that, like other late oppida, the settlement comprised some level of hierarchy. Unlike Entremont's Ville Basse and Le Baou-Roux, however, La Cloche produces a completely non-distributed diagram indicating a restriction of movement around the site (fig 5.37 and fig 5.38).

Communality

Several structures given over to food storage have been identified, and a metal-working area has been suggested (although this last is based on localised finds of tools which were not associated with any formally laid out surface or building). An unusual structure near the summit was associated with various ritual objects and produced no domestic debris, but it is unclear to what extent this was a public or private ritual area (Chabot 1992, 149; 1993, 126-28; 2000, 128). The remains of a warrior statue which bears a ring dated stylistically to the beginning of the third century BC or perhaps slightly earlier appears to have been retained in the *c.* 100 BC settlement, but the identification of a formally laid-out area for its display remains undetermined (Arcelin 2004).

Superstition/Securing of Personal Goods

In addition to the human skulls found in the entrance, which may indicate a ritual head-practice (Chabot 1983, 51), and the remains of a warrior statue strewed along the entrance route (Arcelin 2004, 169), La Cloche had produced copious small objects, many of which appear to be talismans. Pieces of worked coral, certain types of shell, bone or fossil, all of which are abundant at La Cloche, are traditionally preferred material for amulets. Fossils in particular are cited as being endowed with supernatural qualities in European pre- and proto-history (Bourdier 1967, 272-274). Belemnite fossils were found in more than half the houses (Chabot 2004, 169-79).

Nine keys, five bolt runners, three latches for securing house doors and twenty-two door hooks, as well as several locks, keys and hinges from caskets (Chabot 2004, 273-74). Apart from these more formal modes of securing personal belongings, small caches of coins were

found bundled into cavities in at least two house walls and two rings were hidden between bricks in two separate houses (Chabot 1979; 2004, 19, 169). Upturned amphorae, still containing coins, were found scattered in the streets (Chabot 1990, 120). Despite fairly large quantity of residual second century material on the site, only one latch in this collection dates to the first phase of settlement (Chabot 2004, 74).

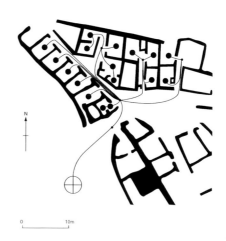

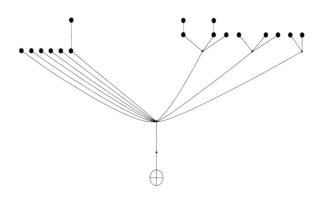

Fig 5.37 (Top) La Cloche: unjustified access diagram. Fig 5.38 (Bottom) La Cloche: justified access diagram. Like Entremont, above, the site demonstrates 'depth', as might be expected in a more complex settlement, however, despite this, La Cloche appears entirely undistributed.

5.15 Patterns of Warfare

Direct Archaeological Evidence for Aggression

The pattern of warfare as seen in direct archaeological evidence (table 5.1) need not be representative of 'real' levels of violence throughout the period. The term 'direct evidence for aggression', it will be remembered, is used here to include buildings demolished, *in situ* objects smashed, and finds of scattered projectiles, possibly overlain by a widespread burnt layer. In other words, the kinds of evidence interpreted as indicating an episode of assault, as described in most of the archaeological reports

6th century	5th century	4th century	3rd century	2nd century	1st century
Tamaris	L'Arquet	Roquepertuse	Roquepertuse	Saint-Blaise x2	Entremont
	L'Île	L'Île		L'Île	Glanum
				Baou-Roux x2	La Cloche
				Teste-Nègre	
				Saint-Marcel	
				Verduron	
				Constantine	
				Glanum	
				Pierredon	
				Entremont	
				Vaison	
				Buffe Arnaud	
				Castellan	

Table 5.1 shows direct evidence for aggression throughout the Iron Age, revealing what appears to be a general spread of warfare on enclosed settlements and major episodes in the second century BC

and syntheses of enclosed, nucleated sites in the region. Direct evidence for aggression, therefore, is less likely to be present in small-scale and/or dispersed agglomerations, which are sometimes identified in this region by such ephemeral evidence as pottery scatters or hearths. Neither can this kind of evidence be expected to identify conflict outwith settlement walls.

Between the sixth and third centuries BC, there appear to be sporadic examples of direct evidence for aggression, particularly around the coastal headlands and lagoon shores, to the west and south of the Étang de Berre. The destruction of the ritual sculpture and portico at Roquepertuse around 250 BC is not included with these, as no other, coeval evidence for assault was identified (an alternative interpretation for this event is suggested in the following chapters). After the third century, there appear to be two concentrations of direct evidence for aggression. The second century BC begins and ends with two striking concentrations of direct evidence for aggression, the first in around 200 BC, the second around 125 BC. The first century BC saw a lessening of such evidence, although the scale and historical specificity of the second assault on Entremont (*c.* 90 BC) has, perhaps, given this particular episode of violence, more prominence than others.

Indirect Evidence for Aggression

The ability to ascertain levels of perceived threat at any given time and place, may suggest levels of conflict among more fragmented populations. The examination of indirect forms of archaeological evidence, for example,

structural defences or weaponry caches, may indicate communities which perceived themselves to be under threat, and made provision to protect themselves. This may be a particularly important indicator for the earlier phase of proto-urbanisation, in the sixth and fifth centuries, which has produced the greatest evidence for dispersed populations. Indirect evidence for aggression here comprises the size and positioning of fortifications and the use of natural topography in the defence of the site. Quantifying approximate, relative levels of defensive intention is fairly straightforward as regards the substance of fortifications. Rampart mass is presented here as a 1 to 4 scale based on the width of the curtain wall. Outworks and other systems of defence are also taken into account (see Appendix 2) but, for clarity, only the simplified version is presented here.

The assessment of 'utilisation of topographical defence' in the construction of defensive systems is more subjective. It is intended to designate an intention to protect the site rather than simply the potential for the site to be defensible. For example, the most substantial section of rampart at the settlement of Notre-Dame-du-Pitié runs along the edge of a cliff around 90m high. While some form of rampart or wall might be expected to protect the occupants from a fatal drop, the 5m thick rampart and 8m wide tower in this area seem superfluous to the protection of the settlement. Conversely, the inhabitants of other sites, such as L'Arquet, were subject to certain hardships by their choice of location; and their fortifications, while substantially smaller, utilise the topography more strategically.

Rampart dimensions	Topographical defence	Total
1-2m (small) = 1	Limited natural defence = 1	**1-2 = little intention to protect**
2-3m (medium) = 2	Poorly used natural defence = 2	**3-5 = average intention to protect**
3-4m (large) = 3	Utilised natural defence = 3	**6-7 = serious intention to protect**
4m+ (extra-large) = 4	Utilised natural defence + exposure/ hardship = 4	**8+ = greatest intention to protect**

Site	Date	Rampart	Topography	Total
Saint-Blaise	6th century	4	3	**7**
Tamaris		4	4	**8**
Castellan		1	3	4
Castillon		3	3	**6**
L'Île	5th century	2	4	**6**
L'Arquet		2	4	**6**
Saint-Pierre		1	1	2
Saint-Pierre	4th century	1	2	3
Notre-Dame		1	2	3
Roquepertuse		1	3	4
Teste-Nègre		2	3	5
Roquepertuse	3rd century	2	3	5
Notre-Dame		4	2	**6**
Saint-Blaise	2nd century	4	3	**7**
Baou-Roux		4	3	**7**
Buffe Arnaud		4	3	**7**
Constantine		3	3	**6**
Entremont		1	3	4
Entremont		3	3	**6**
La Cloche		1	3	4
La Cloche	1st century	1	3	4

Table 5.2 Indirect Evidence for Aggression. Indirect evidence for aggression throughout the Iron Age, showing relative levels of defensive intention. This pattern correlates well with patterns of direct evidence for aggression, suggesting that indirect evidence for aggression may be a reliable indicator of warfare

In the construction of fortifications, the desire to create an impressive, high-status monument must often coincide with the need to protect a settlement: in both protection and display, the sheer mass of the structure would play a significant part. It is not suggested here that the two motivations are in any way mutually exclusive. The rampart wall at Saint-Blaise, for example, is carefully faced so as to be aesthetically pleasing, but special care was also taken in the construction of towers and outworks around the most vulnerable points of the settlement (see Bouloumié 1990, 33-34). It is also necessary to consider the relationship between social complexity and rampart size, as more centralised political control would, naturally, enable the commissioning of large-scale monuments; and fluctuations in consolidation and fragmentation of settlement hierarchies must also be taken into account. Nevertheless, it is interesting that indications of a serious intention to protect these settlements (estimated from a combination of rampart

mass, strategic emplacement and a willingness to endure certain physical hardships) seems to exclude settlements which had impressive fortifications but which did not produce evidence for direct aggression, such as Saint-Pierre and Notre-Dame. Conversely, combinations of topographical and strategic defensive systems appear to occur most strongly in the period of emerging proto-urbanisation in the sixth century, and in the second century BC. The relationship between the ancient texts and archaeological evidence is discussed in more detail below, but the above pattern of perceived threat felt by the indigenous populations does not appear to correlate with the documentary accounts of conflict in the region. The phase in which the builders of oppida appear to have been least concerned with defensibility, coincides with the fourth and third centuries. This encompasses the siege of Massalia (390 BC), and the time of 'unrest' between the two groups; as well as the period traditionally associated with the major period of Phocaean expansions throughout the region (table 5.9).

Socialisation for Mistrust

Fragmentation

As can be seen from table 5.3, even the larger households of the earlier period fall below or just within the 22.86 – 40m² range of internal floor-space proposed in this study as representative of a nuclear household. The limitations of determining 'small' households have been outlined briefly above, and the usefulness (or otherwise) of household size, as an indicator of embedded warfare will be discussed further in chapter eight. For now, house sizes are considered as relative quantities. The smaller households tend to group between the fourth and early second centuries BC, the fourth- and third-century households appearing particularly small, with households of 10m² and less.

Access Analysis

Access analysis was carried out on those settlements with sufficiently complete site plans, in an attempt to gauge confidence in social relationships beyond immediate family members. It was suggested that a breakdown of trust within communities might be shown in non-distributed diagrams (i.e. diagrams with few horizontal connections) which indicate restrictions of movement between households. The results of this analysis showed an almost universal lack of distribution, or connection between households, partly due to restrictions of movement inherent in the indigenous building tradition of the period. For this reason, again, the results of this analysis are assessed comparatively, to evaluate relative restrictions of access between individual settlements and throughout the period (table 5.4). Restricted access is presented here on a scale of 1–3, where 1 represents a completely non-distributed diagram, 2 indicates a minimally distributed layout (perhaps suggesting that the lack of movement is incidental to the form of layout and not a deliberate choice) and 3 signifies a more distributed settlement layout (i.e. more than usual ease of movement).

Restrictions of movement around settlements do not appear to change appreciably until the period of the complex settlements of Entremont Ville Basse and Le Baou-Roux. Interestingly, however, there seems to be a constriction of access in some settlements (e.g. the deliberate blocking of streets) prior to episodes of destruction, or, in the case of Saint-Pierre, at a period of concentrated violence in the region. The most compelling results come from L'Île, where the more refined chronological sequence allows a more precise consideration of each change in access, in relation to other potential indicators of aggression. The period of exceptional ease of access, and the period of the largest average households (17.30m²) are both seen during an apparently peaceful interlude of several generations, between *c.* 430 and *c.* 360 BC. Subsequent to the *c.* 360 BC destruction, access was more restricted and households reduced to 10.30m². Ease of movement

constricted further prior to *c.* 200 BC, a period of concentrated warfare in the region.

Communality

Specifically-constructed gathering places, such as public squares, or 'hypostyle' halls (rooms dedicated to communal drinking rituals), rarely occur until the period of the more complex oppida of the mid- to late-second century BC. This analysis is concerned with communality as a reflection of trust-relationships outside the immediate family, therefore, in addition to such formally-constructed public spaces, this section also observes the appearance (or otherwise) of less-formal, shared-spaces. These may be storehouses, food-processing areas or craftworking-areas, and comprise adapted domestic structures or simply un-built space. They may have been used by a few rather than large sections of the community, but may represent, nevertheless, specially-designated, shared space. In addition, it should be noted that some sites produced sculptural fragments which pre-dated the settlements on which they were found. These may indicate that the settlements had once been sites of public ritual; however, as discussed in the following chapter, the exact location and manner of display of these statues cannot be definitely established (table 5.5).

The Martigues area yielded very few public spaces, however, it is difficult to ascertain to what extent these differences in communality appear to be geographical or temporal. Early sanctuaries may be inferred from fragments of fifth-century warrior statues found at the sites of later settlements in the Massaliote hinterland, however, these have left no remains. There are more communal spaces in the Marseille region in the fourth and, particularly, the third centuries BC. These are generally informal shared spaces, such as the food-storage and food-processing areas in the late third-century phases of settlement at Le Baou-Roux, Teste-Nègre and Roquepertuse. The increase in such areas in the third century may indicate an increasing level of communality, however, this is also the period of the smallest houses, with average floor-spaces such as 8m² at Teste-Nègre and 11m² at Notre-Dame. This may indicate the greater need for storage and work-areas in settlements where households have little floor-space; perhaps making houses at the lowest end of the size range unsuitable indicators of mistrust. In any event, communal spaces cannot be said to be avoided during this period, as public ritual-places can be inferred (if not precisely located) from finds of fifth- and third-century sculpture (see chapter six).

The apparent increase in both utilitarian and ritual public spaces in the third century is probably a result of the increasingly centralised control of resources during the period of consolidation in the indigenous settlements around Marseille during the fourth and third centuries BC. This is not to say that these spaces do not reflect increases in trust relationships among the resident communities. Many of these shared spaces are no longer in use around the beginning of the second century. At Le

Baou-Roux, for example, the communal storage and artisanal areas disappeared, streets were narrowed and, by the end of the third century, blocked off. At Entremont, too, a seemingly substantial, third-century sanctuary (indicated by numerous, ritualistic carvings and the fragments of around twenty statues), appears to make way for the Ville Haute, in the early-second century. This first phase of enclosed settlement on the site comprised small, densely-grouped houses, narrow streets and, apparently, no public areas. These transformations may indicate a breakdown of trust relationships in communities outwith immediate family units.

The decline of communality certainly appears to coincide with the concentrated period of direct evidence for aggression around 200 BC.

Formal, specially-constructed, public meeting and ritual spaces became more usual at later second-century settlements, such as Entremont's Ville Basse. Such sites represent the period of greatest social organisation and political centralisation among the indigenous populations of the region, and monumental public architecture must be seen as a manifestation of this.

	6th C	5th C	4th C	3rd C	Early 2nd C	Late 2nd C	Early 1st C
Tamaris							
Saint-Pierre	20m^2	20m^2	8m^2	8m^2			
L'Arquet		16m^2					
L'Île		13m^2	17.3m^2	10.3m^2	13m^2		
Notre-Dame			11m2	11m2	11m^2		
Roquepertuse		24m^2		21.5m^2			
Teste-Nègre				8m^2			
Baou-Roux					25m^2	25m^2	
Entremont					10m^2	49m^2	49m^2
La Cloche						13.5m^2	40m^2

Table 5.3 Comparative House Sizes throughout the Iron Age. The size range suggested to denote a 'small' household was 22.86-40m^2 or less. In terms of *relative* size, small houses can be considered those of 15m^2 or less, medium houses fall between 16 and 19m^2 and the largest households (in bold) are 20m^2 and over

Site	Date	Ease of Movement	Other Factors
Tamaris	600 BC – 550 BC	3	
Tamaris	Prior to 550 BC destruction	3	blocked streets
Saint-Pierre	550 BC – AD 100	**1**	
Saint-Pierre	c 200 BC – AD 100	**1**	blocked streets
L'Île 2b	2nd phase of rebuilding after 430 BC destruction	3	
L'Île 2c	Immediately prior to 360 BC destruction	3	
L'Île 3a	Immediately after c 360 BC destruction	2	
L'Île 3b/4	Immediately after 360 – 200 BC destruction	**1**	
Notre Dame	325 BC – 260 BC	**1**	Wide streets
Baou-Roux	200 BC – Prior to 125 BC	3	
Baou-Roux	Prior to 125 BC destruction	3	blocked streets
Entremont	175 BC – 150 BC destruction	**1**	
Entremont	150 BC – 90 BC destruction	3	Wide streets
La Cloche	100 BC – 50 BC destruction	2	

Table 5.4 Socialisation for Mistrust: Ease of Movement around Settlements.
Greatest restriction of movement around site (no interlinking streets) = 1(in bold).
No deliberate restrictions of movement (one inter-linking street) = 2.
Ease of movement (more than one inter-linking street) = 3.

	6th Century	5th Century	4th Century	3rd Century	Early 2nd Century	Late 2nd Century	Early 1st Century
Saint-Blaise	-	-	-	-	-	present	
Tamaris	absent						
Saint-Pierre	sanctuary	-	-	-	-	-	-
L'Arquet		absent	absent				
L'Île		absent	absent	absent			
Notre-Dame			-	present			
Roquepertuse		present	present	sanctuary	shared space		
Glanum				sanctuary	present	present	
Teste-Nègre				shared space			
Baou-Roux				shared space	absent	absent	
Entremont				sanctuary	-	present	
La Cloche				sanctuary			present

Table 5.5 Public Spaces throughout the Iron Age.
Absence of communal space (possible indication of mistrust within the community) = absent (bold).
Informal shared space, storehouses, open areas with evidence of craftworking etc. = shared space.
Formal and specifically structured public spaces, hypostyle rooms etc. = present.
Ritual space may be inferred on sites which have produced statuary fragments = sanctuaries.

Superstition

It is suggested here that superstition (as an indication of fear and mistrust within communities) may be inferred by finds of personal amulets and/or household or domestic ritual (as distinct from public ritual or religious foci. The presence of superstition in the analysed settlements is described in the table (table 5.6) on a rising scale of present, frequent or abundant.

The extent to which superstition is present in settlements appears to be drawn along geographical rather than temporal lines, with the Martigues region producing abundant and widespread evidence for personal amulets and household devotions in all periods. Many of the domestic rituals in evidence in Martigues, such as depositions of animals or parts of animals in house floors, are more usually associated with ritual practices seen in the Languedoc, while others, in particular the deposits of small pierced objects, occur all over southern France. East of the Étang de Berre the selected sites produced evidence for personal amulets only from the second century BC at Le Baou-Roux and Entremont Ville Haute. The most notable exception to this dichotomy is the first-century BC site of La Cloche, which produced profuse evidence for personal talismans in the majority of domestic units.

Securing of Personal Goods

Evidence for the securing of personal belongings (table 5.7) such as caches of coins and, in particular, finds of keys and locks, occur most frequently at the later, complex oppida. This may, in part, reflect the more centralised control of communal resources and personal wealth at these more hierarchical settlements. The exceptions to this are the early phases of settlement at L'Île, Entremont Ville Haute and La Cloche, which produced evidence for caches of coins and valuables hidden in house-walls (Willaume 1993; Chabot 2004, 19 and 169; Chausserie-Laprée 2005, 220).Evidence for securing personal goods may be the most difficult indicator of mistrust to establish. Hidden valuables must often have been removed, either as plunder or in the retrieval of personal items by the occupants of a settlement following an assault. In addition, many sites (e.g. Tamaris and L'Arquet) have produced evidence of a brief re-occupation between an assault and the eventual abandonment of the settlements. The discovery of such caches at some of the most thoroughly investigated settlements in southern France (i.e. L'Île and Entremont) suggests a more widespread phenomenon, as yet unidentified at other sites in the region.

		6th C	5th C	4th C	3rd C	Early 2nd C	Late 2nd C	Early 1st C
Martigues	**Tamaris**	-						
	Saint-Pierre	3	3	3	3	3	3	3
	L'Arquet			3				
	L'Île		2	2	2			
Massaliote hinterland	**Notre-Dame**			-	-			
	Roquepertuse				-			
	Teste-Nègre				-			
	Baou-Roux					1	1	
	Entremont					1	-	
	La Cloche							3

Table 5.6 Superstition throughout the Iron Age. Superstition is inferred here from the presence of amulets and/or ritual material in domestic contexts. The greatest occurrence of this appears to be associated with Saint-Pierre, one of the most stable settlements of the region. Over and above the Embers' findings, the almost ubiquitous association of superstition with dangerous or unpredictable situations, this may highlight the difficulty (or impossibility) to identify superstition on an individual or domestic level in the archaeological record
Present = 1.
Frequent = 2.
Abundant = 3.

	6th C	5th C	4th C	3rd C	Early 2nd C	Late 2nd C	Early 1st C
Tamaris	-						
Saint-Pierre	-	-	-	-	-	-	-
L'Arquet		-	-				
L'Île	2	2	2				
Notre-Dame			-	-			
Roquepertuse				-			
Teste-Nègre				-			
Baou-Roux					-		
Entremont					3	3	
La Cloche							3

Table 5.7 Evidence for the Securing of Personal Goods. The securing of personal goods is proposed here to be represented by the presence of locks and/or keys, and caches of coins and valuables. The results of this analysis are quantified as follows:
Presence of evidence for the securing of personal goods = 1.
Frequent finds of evidence for the securing of personal goods = 2.
Abundant finds of evidence for the securing of personal goods = 3 (bold).

5.16 Summary

Patterns of Warfare

This section compares the patterns of warfare constructed from direct evidence for aggression, indirect evidence for aggression, and some suggested manifestations of socialisation for mistrust, in the selected settlements (table 5.8). In order to better understand the relationship between warfare and society (discussed further in chapter eight), the chronological sequence must be adjusted slightly from the broad, two-century divisions of urban development, outlined in the early part of this chapter. Accordingly, it is necessary to note that, with the exception of Roquepertuse, all the direct evidence for aggression in the fourth and third centuries BC takes place in the early-fourth century BC. These assaults took

place in the same geographical area as the early 'proto-urban' period warfare, and involved the same settlement (L'Île), which shows no signs of substantial organisational or economic change at that time. This may suggest that the early fourth-century assault at L'Île was part of the same dynamic as the early episodes of conflict in Martigues.

Similarly, the changes in indigenous settlement organisation, from small, dispersed communities to large, complex centres, occurred in the course of the mid- to late-second century BC. Consequently, the two great concentrations of direct evidence for aggression in *c.* 200 BC and *c.* 125 BC took place against different socio-political backgrounds and are, therefore, implicated in very different social and political processes. For this reason, the appraisal of patterns of warfare, below, considers 'early warfare' to indicate conflict between the sixth and early-fourth centuries BC, and differentiates between the early-second century BC and the period of complex oppida in the late-second and early-first centuries BC.

Direct Evidence

Direct evidence for aggression comprised one or more indications for a direct assault, including destruction-levels, fire-damage, projectile-scatters and evidence for violent death. This pattern indicated sporadic episodes of violence in the early centuries of urban development, in the region of Martigues, where such settlement initially occurred. The 'intermediate period', which saw the consolidation of urban form and function, east of the Étang de Berre, appears to show an apparent absence of assault on the new, upland oppida, however, this period ends in a concentrated upsurge of violence in around 200 BC. The large, complex oppida of the following period also witnessed a concentrated period of warfare linked, through known Roman historical sources, to the campaigns against the Saluvii at the end of the second century BC.

Indirect Evidence

Indirect evidence for aggression considered the size and positioning of fortifications, the use of natural topography, and the willingness to endure certain natural hardships, in the interests of defence. To a great extent, indirect evidence for aggression supports the patterns of warfare as they appear in direct evidence for assault shown above. The majority of the early (sixth- to early-fifth centuries BC) sites, and all the second-century sites demonstrated 'serious intention to protect' the settlements. Sites of the third and fourth centuries BC, however, demonstrated, 'little intention to protect settlements'.

Socialisation for Mistrust

Evidence for socialisation for mistrust comprised a range of data proposed to reflect the breakdown of trust relationships beyond the immediate family within the analysed settlements. These comprised evidence for fragmentation (small households, restrictions of access around settlements, little communality), the securing of personal belongings (keys, locks and caches of valuables), and superstition (amulets, domestic ritual). The analysis of these data from the selected sites produced mixed results. Many of the criteria suggested to demonstrate the securing of personal goods was ambiguous. Caches of coins and valuables were discovered at sites such as L'Île and Entremont (which had undergone repeated assaults) and at La Cloche (a final bastion of indigenous settlement in the Roman province). Finds of locks and keys, however, may have been associated with changing levels of socio-political complexity as much as indicating levels of mistrust within communities. Evidence for superstition also proved inconclusive. In particular, the frequency of certain domestic rituals in evidence throughout

Martigues, such as depositions of animals beneath house floors, may simply have reflected this region's proximity with the Languedoc, where such practices were common. The identification of fragmentation presented some problems, as the arbitrary, 22.86–40m^2 range formulated to denote 'small' households, was shown to be larger than all but the largest households from complex oppida. Nevertheless, comparative house-sizes as an indication of relative fragmentation, along with evidence of communality and restrictions of access around settlements, produced significant results. Comparatively small households and communal spaces in particular appear to follow the same patterns of warfare revealed in more conventional forms of evidence, such as destruction levels and defensive systems.

	6th–early 4th C	Mid-4th– 3rd C	Later Iron Age Early 2nd C	Late Iron Age 2nd – early 1st C
Direct Evidence for Aggression	some	least	greatest	greatest
Indirect Evidence for Aggression	some	least	greatest	greatest
Fragmentation Household Size	large	smallest	small	largest
Communality Shared Space	some	frequent	none	greatest

Table 5.8 Evidence for aggression in the Lower Rhône Valley.
The direct and indirect evidence for aggression correspond very well throughout the period. These appear to have a negative correlation with the evidence for fragmentation and communality.
Regarding evidence of socialisation for mistrust, the most compelling results derive from evidence for small households and lack of communality.
The very small houses and increased communality in the fourth and third centuries may reflect the need for extra work and storage areas where domestic space is circumscribed (although other interpretations are discussed in chapter eight).
All four forms of evidence appear to correlate well during the early-second century.

5.17 Comparison with Documentary Evidence

The historical framework of events discussed in chapter four (table 4.1) appears to correspond at several points with the patterns of warfare taken from the settlement evidence. There are references to unspecified conflicts between Massalia and the indigenous population in the fifth century BC, perhaps alluding to events surrounding the assaults on L'Île and L'Arquet *c.* 460/440 BC. However, similarly vague reports are also assigned to the third century, for which the settlement evidence produces only one example for direct evidence for aggression, at Roquepertuse.

Conversely, the great concentration of violence at the beginning of the second century in the Massaliote hinterland appears to have gone unrecorded. As outlined in the previous chapter, most of the literary evidence for warfare comes from second-hand accounts of original sources, themselves composed several centuries after the events. It may also be relevant that most of the authors were recording histories of the rise of Rome or the wars with Carthage, to which 'events' in southern France were peripheral to the central story. The only specific event comes from Pompeius Trogus, a native of southern Gaul, who recorded the Gallic attack on Marseille in *c.* 390 BC (Justinus 43.5). This may be relevant to the abandonment and burning of the small, early and unenclosed settlement at Roquepertuse *c.* 400 BC. Otherwise, there is no substantial correlation between the documentary evidence and episodes of warfare derived from the settlement evidence, until the Roman campaigns, initiated by Quintus Optimus in 154 BC and ending with Caesar's wars against Pompey in the first century BC (see tables 5.9 and 5.10).

Chapter four identified several shortcomings in the documentary record, primarily, the bias toward large-scale and externally directed warfare which disregards possible small-scale conflicts which may have taken place between Greeks and Gauls, as well as internecine conflict within the indigenous communities. There is also the consideration that texts referring to early warfare may have been lost in the assault on Rome in 390 BC or during the siege of Massalia in 49 BC. In either case, possible early conflicts and conflicts not involving Roman forces are, for the most part, missing from the historical record. Dominique Garcia (2004, 91) sums up the remarkable concentration of violence around 200 BC in a few lines, stating that at this time in the vicinity of Marseille:

> ...*numerous sites were abandoned, some following a violent destruction (Le Verduron and Teste-Nègre). After this stage, new sites were created...*

A similar lack of detail and analysis regarding the possibility of pre-Roman warfare is also the norm in individual site interpretations. This can be encountered frequently in the *Bilan Scientifique* reports (annual publications of the results of excavations in southern France). In these articles, levels of destruction or burning are alluded to briefly, merely as aids to identifying stratigraphy, or preserving grain or other organic remains (e.g. Bernard 1999, 99-100; Verdin 1991, 111).

Perhaps the most important misconception which can be attributed directly to the written sources, however, is the impression that conflict in the region was, at all times, drawn up along ethnic lines of opposition. Some episodes of warfare after the mid-second century appear to relate to specific campaigns during the Roman intervention (table 5.10). Prior to the Roman campaigns of the mid-second to mid-first centuries (table 5.9), however, there is no such link between archaeological evidence for aggression and any specifically mentioned conflict. Despite this, pre-Roman evidence for warfare has routinely been attributed to the hostile and unstable relationship between the colonising Greeks and their indigenous neighbours. The historical record sees two periods of conflict associated with this period of the Phocaean colonisation (see chapter four), essentially summed up by Justinus and Strabo as follows:

> *Having defeated their enemies, they established many colonies in the conquered lands (Justinus 43.3.13).*

> *They followed their natural inclination for the sea, but, later, their valour enabled them to take in some of the surrounding plains, thanks to the same military strength by which they founded their cities (Strabo 4.1.5).*

From this it is often inferred that evidence for warfare in the region (when noted) was associated with the foundation of Massalia around 600 BC and the expansion of Greek colonies in the fourth and third centuries BC.

One of the more rigorous readings of a specific, archaeologically-attested conflict comes from Boissinot's interpretation of the third-century assault on Roquepertuse, which admits that the assailants cannot be identified. More often, however, the assumption that early hostilities were drawn up along Greek-Gaulish ethnic lines is simply assumed. Garcia (2004, 85), for example, interprets changes in the apparent fragmentation of indigenous settlement in the fourth century BC, in a general and rather indefinite manner, as the subjugation or forced retreat of the Gauls in the face of Greek expansion. While Jean Chausserie-Laprée explicitly offers an even less nuanced interpretation for the series of siege assaults concentrated around 200 BC:

> *At the end of the Hellenistic period a renewal of nucleated settlement is shown. It occurred after a new phase of political and military crisis which, around 200 – 190 BC, saw Greeks and Gauls in opposition once more around the Étang de Berre (2005, 47).*

Site	Destructions	Historical Warfare
		154 BC Q. Optimus's campaign (Strabo)
Constantine	125 BC	125 BC M. F. Flaccus' campaign (Livy)
St. Marcel	125 BC	125 BC M. F. Flaccus' campaign (Livy)
Saint Blaise	125 BC	125 BC M. F. Flaccus' campaign (Livy)
Buffe Arnaud	125 BC	125 BC M. F. Flaccus' campaign (Livy)
Le Baou Roux	125 BC	125 BC M. F. Flaccus' campaign (Livy)
		123 BC permanent Roman garrison at Aix
		106 BC Caepio's Campaign
Entremont	90 BC	102 BC Marius's campaign
Glanum	90 BC	
La Cloche	c50 BC (Rome)	49 BC Siege of Massalia by Caesar

Table 5.9 Some possible correlations between settlement evidence for warfare and historical accounts of warfare before the Roman intervention

Site	Destructions	Historical Warfare
Entremont	150 BC	154 BC Q. Optimus's campaign (Strabo)
Constantine	125 BC	125 BC M. F. Flaccus' campaign (Livy)
St. Marcel	125 BC	125 BC M. F. Flaccus' campaign (Livy)
Saint Blaise	125 BC	125 BC M. F. Flaccus' campaign (Livy)
Buffe Arnaud	125 BC	125 BC M. F. Flaccus' campaign (Livy)
Le Baou Roux	125 BC	125 BC M. F. Flaccus' campaign (Livy)
		123 BC permanent Roman garrison at Aix
		106 BC Caepio's Campaign
Entremont	90 BC	102 BC Marius's campaign
Glanum	90 BC	
La Cloche	c50 BC (Rome)	49 BC Siege of Massalia by Caesar

Table 5.10 Some possible correlations between settlement evidence for warfare and historical accounts of warfare after the Roman intervention

5.18 Identification of Assailants through Finds of Weaponry

While the origin of this misapprehension lies in an unrevised, early reliance on the documentary sources, physical evidence for this interpretation is based entirely on the supposition that scatters of Greek weaponry from destruction layers indicate a Greek assault (e.g. Chausserie-Laprée 2005, 97).

Specialist Weaponry

In order to critically assess the possible ethnic identities of those participating in conflict in the region, it is necessary to re-evaluate current interpretations of finds of both specialist and improvised weaponry on indigenous sites. In the Provençal Iron Age specialist weapons such as swords are rarely associated with either settlements or burials. Throughout the proto-historic period, weaponry is overwhelmingly evidenced by projectiles. These are generally catapult missiles or 'boulet' of stone or terra cotta; metal points of Greek manufacture: usually bronze arrowheads or iron javelin and spearheads; or Roman pilum points and spear butts. According to current interpretations, these objects, when found in the indigenous milieu, are usually considered weaponry only when they are directly implicated in a siege assault, that is to say, when they are found scattered around the site and/or included in destruction levels. One notable exception is the interpretation of a stash of catapult missiles found in a workshop in Entremont around the time of the final assault on the oppidum (Benoît 1975, 231). Greek projectile points, and catapult missiles of either Greek or Gaulish origin, have been found in formal, collected stashes or in domestic structures in many indigenous settlements; however, in this context they are considered to be intended for hunting.

As a consequence, these missiles, and the manner in which they are deposited, are often used to identify the ethnic origin of the perpetrators of a military assault. This assumption is strongly implied in the 'Greek assault' / 'Roman assault' interpretations used to identify different episodes of siege warfare in the region; and overtly stated by Chausserie-Laprée when he describes how bronze arrowheads found around the walls at Tamaris, identified the Greeks as the assailants, but that those found in houses in the village of Saint-Pierre were bought or traded from the Phocaeans for hunting (2005, 192). At most it is considered that indigenous assailants may have carried out an attack instigated and backed by Massalia (*ibid.*, 97).

The suggestion made by Christian Goudineau (1983) that the Massaliotes were trading war technologies to the indigenous populations of southern France is borne out by modern analysis of stone-cutting techniques (Bessac 1980) which reveal indigenous masonry on Greek-style fortification early in the proto-historic period. Goudineau cites Strabo's account (VI, 1, 5) of the Massaliotes as the educators of the Gauls and (in the same passage) as experts in the instruments that are useful for the purposes

of navigation and of sieges (1984, 82-83). It seems a straightforward enough conclusion that, if the indigenous populations of Provence adopted defensive technologies as complex and work-intensive as the ramparts enclosing the early settlements of the sixth and fifth centuries, then they very likely acquired or learnt to make the offensive missiles, the spears and arrows, that opposed them.

Improvised Weaponry

Before leaving the subject of weaponry in indigenous contexts, it is apposite to mention the use of improvised missiles. Lead, stone and terra cotta fishing weights have been recovered from many indigenous sites, in particular those along the littoral and lagoon coasts in the early proto-historic period. They have also been found in concentrations around the foot of the Hellenistic tower of Saint-Blaise, and stashed at strategic points around the rampart walls (Chausserie-Laprée 2005, 191) suggesting that they were used as sling missiles in defence of the settlement. Although these and other such items must be considered primarily as tools, they highlight the problem of designating certain less ambiguous items 'tools' or 'weapons' according to the ethnic milieu in which they are found.

5.19 Conclusion

The above résumé of patterns of conflict and the physical evidence defining them raises questions about the usefulness of weaponry as a tool for ascertaining who is fighting whom in a scenario in which the Greeks are essentially arms-traders, and the Gauls their willing customers. Recognising the causes and consequences of these episodes of violence in terms of the mechanisms driving periods of intense conflict and their repercussions on political complexity and social organisation, depends to a great extent on whether the archaeological evidence represents cross-cultural or internecine violence. Current study takes the position that warfare in the region was carried out across inter-ethnic lines. Despite the weight of archaeological and historical tradition, however, there is no sustainable archaeological evidence for these conclusions.

6.1 Introduction

The extraordinary and powerful lapidary collections from Entremont and Roquepertuse have made the iconography of the southern French Iron Age one of the most widely known aspects of its archaeology. Throughout the Lower Rhône Valley, settlements of the Second Iron Age have produced a range of distinctive stone sculptural remains apparently preoccupied with themes of warfare and headhunting. The anthropomorphic figures are usually (though not always) male, and in armour. Some of these 'warrior statues', notably those from Entremont, hold carved, severed heads under their hands or in their laps. Engraved and painted friezes and blocks of stone have also been discovered, bearing similar martial and hieratic images, including human heads, horses, and figures involved in warlike or ritualistic activities. In addition to these are a series of stone pillars, created to stand upright, as stelae, or sometimes mortised to receive a cross-beam of wood or stone to form a portico. These pillars were carved with representations of human heads, or had niches cut into them to receive actual heads or skulls (fig 6.1).

6.2 Aims

As with the study of conflict in the region, the treatment of the iconography of southern France has been dogged by a culture-historical paradigm that other areas of research have long left behind. During the nineteenth and early twentieth centuries, when much of this material was first discovered, the warlike nature of many of these images was seen as supporting the violent and barbaric image of the native Celto-Ligurian tribes described in the ancient texts. In particular, the iconography pertaining to head-related ritual practices is still often seized upon to illustrate Poseidonius's famous passage describing headhunting practices among the southern Gauls. Moreover, from the earliest investigations (e.g. Gérin-Ricard 1927; Benoît, 1955, 8) to the present day (e.g. Salviat 1989; Rapin 2003, 244) research has been preoccupied with establishing the degrees of Hellenisation versus indigenous idiom which influenced the creation of these sculptures; a debate which signally fails to consider the objects as cultural expressions within their own milieu.

This chapter presents a more contextual account of the iconography, and attempts to determine whether the warlike imagery apparent in this tradition, truly represented a conflicted society. The various kinds of iconography will be considered thematically, with reference to the archaeological and historical circumstances of their known creation, deposition and display.

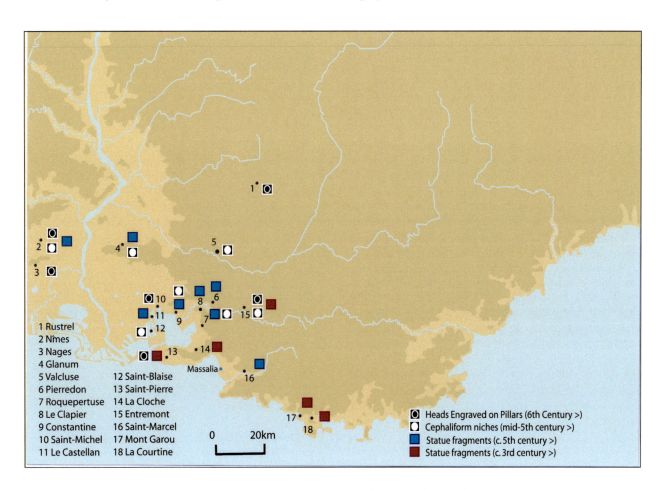

Fig 6.1 Distribution of representations of têtes coupées and warrior statues

Firstly, depictions of human heads will be examined. Disembodied, human heads are, arguably, the premier motif of the southern French Iron Age, and appear in various forms, engraved on stelae and architectural elements, or carved in three-dimensions in sculptural compositions. Secondly, anthropomorphic representations will be considered. As with the head-motif, engraved or sculpted human figures are known in southern France from the Early Iron Age. While some female figures are known, human representations are overwhelmingly 'warriors' or at least, figures in armour, sitting, riding or carrying out ritualistic activities. Finally, various iconographic representations (usually engravings) are examined. These vary in form, but usually comprise depictions of ritual practices or symbolic motifs.

It should be noted that there is much overlap within this division of the iconography, which is presented here simply as an arbitrary framework within which to discuss this material. At the conclusion of this chapter, the different forms of indigenous iconography (human heads, anthropomorphic representations and ritualistic depictions and motifs) will be examined in tandem. In particular, changes and continuities will be noted, as regards their morphology, the location and manner of their display, and how these may have changed throughout the period. It may then be possible to ascertain how these images of warriors and heads relate to the history of warfare in southern France, and to the patterns of violence compiled from the settlement evidence in the previous chapter. Central to this discussion is the chronology of the various types of sculptural remains; a matter which is complicated by a range of difficulties (outlined below) which recent work in this area has begun to address.

6.3 Chronology

The chronological framework which encompasses all these lapidary remains is still in the process of re-evaluation. The anthropomorphic statuary, best known through the large collections from Entremont and Roquepertuse, were traditionally felt to represent the adoption of Hellenistic sculptural techniques (e.g. Benoît 1969; Salviat 1989). The naturalistic conception of these statues, and the style of their technical execution, so greatly resembled the conventions of Greek plastic art, that, until recently, they were assumed to be an advanced, and therefore, late, form of this tradition (e.g. Salviat 1993, 238). Moreover, most of the fragments had been recovered in or close to sites of the later Second Iron Age (Arcelin 1992; Guillet et al. 1992). Consequently the creation of these figures was simply dated to the occupation of the sites (second century BC for the Entremont collections, third/second century BC for the statues from Roquepertuse and Glanum etc.) (Arcelin and Rapin 2003, 185).

Naturally this has long been considered an unsatisfactory dating method (*ibid.*), but several fundamental obstacles lay in the way of establishing a more accurate chronology for these works. Firstly, virtually none of these elements

had been found in a primary context. As described in more detail below, there appears to have been a widespread tradition of curating and re-using iconography throughout the Iron Age. Fragments of statuary and other stone iconography have been found re-employed within later structures and some statues are now believed to have been moved to their final settings, in later Iron Age settlements, two or three centuries after their creation (Arcelin et al. 1992). In addition, the vast majority of these objects had been broken in antiquity; often during the episodes of violent destruction in the second and first centuries BC described in the last chapter. Finally, these problems have been confounded by the often haphazard nature of the sculptures' ultimate retrieval. Few pieces come from modern excavations, and stratigraphic recording of their recovery is often ambiguous or simply non-existent.

In recent years, new research has allowed a revision of the traditional chronology; particularly as regards the anthropomorphic statuary. Notable among these new data are the warrior statue found reused as building material at Lattes (Dietler and Py 2003) and research into Iron Age weaponry carried out by Lejars (1994) and Rapin (1999). This combination of stratigraphic, artefactual and stylistic dating, dates several of these warrior statues to the end of the First Iron Age, and indicates that, (in differing forms), this tradition continued to be employed throughout the period. Similarly, a chronology starting from the end of the Late Bronze Age has been suggested for pillars engraved with human heads, based on the stylistic cross-referencing of zoomorphic figures which sometimes accompany the main head motif (Arcelin and Rapin 2003, 188-89). As with the statuary, stylistic differences in later settings suggest that the image of carved heads and/or human crania on pillars and porticoes constituted an ongoing and developing iconographic tradition (*ibid.*, 191).

This new chronology (Arcelin and Rapin 2003) does not pretend to provide precise dates for individual examples of iconography. Many pieces in the following discussion must be assigned simply to the First or Second Iron Age where the data to suggest a more specific date is not available. As with the examination of the settlement evidence, therefore, this chapter begins with the caveat that all the proposed dates are approximate. Nevertheless, it should be possible to trace broad changes in theme or emphasis throughout the period, and compare what might be termed 'surrogate' (Osgood 1998) or 'rhetorical' (Armit et al. 2006) violence, to the patterns of warfare documented in the previous chapters.

6.4 The Iconography

6.5 Têtes Coupées

Human heads or faces are a ubiquitous motif in the Iron Age iconography of Europe, recurring frequently in La Tène artwork (Jacobsthal 1941, 303; Megaw 2003, 269). The southern French manifestation of this image is the disembodied or severed human heads, described as *'têtes coupées'* (Salviat 1993, 209). Carvings of heads laid out in vertical rows on wood and stone have been discovered occasionally further north (similar objects have been recovered from the Seine for example (Ross 1986, 117)), and west (in Sant Martí Sarroca near Barcelona (Lenerz-de Wilde 1995, 545)). In southern France, however, this image was a staple motif. Human heads occurred as engravings on fragments of architectural masonry, stone pillars and lintels, and in three-dimensional sculptural compositions, from indigenous settlements around the Lower Rhône Valley.

Pillars and Lintels Engraved with Heads

Human heads engraved in rows occur at several sites in southern France, though most were recovered from settlements around the Lower Rhône. The earliest-known engravings of such heads take the form of *bétyles*, crudely-shaped upright stones (fig 6.2) or more carefully-dressed pillars (Arcelin and Rapin 2003, 186-89) (fig 6.6). In some compositions, the shallowly engraved heads are identical and schematic, with eyes and nose indicated by a single, unbroken line and mouths not always included (Benoît 1964). The heads seem never to have been represented singly, but appeared in pairs, or, more usually in rows, either horizontally on a lintel or vertically on a pillar (Arcelin and Rapin 2003, 190). A frequent characteristic of the vertical rows of heads is the *tête-bêche* (*ibid.* 188) or inverted, mirror-imaging of heads in the composition (fig 6.7). Most of these pillars and lintels engraved with heads are known from secondary contexts, so that a precise dating is impossible.

The inclusion of a horse motif on a *bétyle* from Var, which corresponds stylistically with engravings from Mouriès and Glanum (fig 6.3), suggests that the beginnings of this tradition can be assigned to the transition between the Late Bronze and Early Iron Ages (Arcelin and Rapin 2003, 189). More developed forms, such as the pillar from Entremont (fig 6.6) are believed to date to the end of the seventh or beginning of the sixth century (*ibid.* 192). There is no evidence that such pillars were still being created during the Second Iron Age, however, they continued to be used, incorporated into third- and second-century structures. The pillar from Entremont, for example, was found re-used as a stylobyte in the façade of a second-century public building (Salviat 1993, 209). A similarly engraved stone stele was discovered, re-used as building material in a second-century BC house at Entremont. Known as the *bloc aux épis*, two sides bear rows of like those dated to the late-seventh/early-sixth centuries (fig 6.4) (Arcelin and Rapin 2003, 188), while the surface facing into the interior of the house was engraved with ears (*épis*) of wheat (fig 6.5).

Heads engraved in heavier relief are believed to be later (Arcelin and Rapin 2003, 192). These have more detailed and naturalistic, though still undifferentiated, facial features (fig 6.8). Sometimes these high-relief rows of heads incorporate other motifs such as cephaliform niches (see below). The lintel from Glanum (fig 6.9) which

alternates carved heads with running horses is dated, stylistically, to the Second Iron Age (Py 1990).

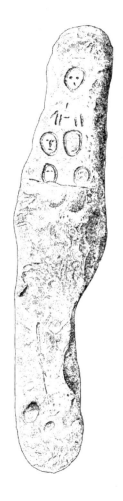

Fig 6.2 (left) Bétyle from Var bearing a horse-motif stylistically comparable to engravings from Mouriès (fig 6.3, below) and Glanum which date to around the eighth and seventh centuries BC.

These early occurrences of rows of têtes coupées appear simplistically delineated and were pecked rather than incised. (Drawing from Arcelin and Rapin 2003, 189; photo: Ian Armit).

Fig 6.4a and fig 6.4b (above left) The so-called bloc aux épis from Entremont. On two sides the stone carries crudely-rendered versions of head motifs found on pillars and standing stones throughout the region (from Arcelin and Rapin 2003, 188).

Fig 6.5 (above right) The face of the bloc aux épis which gives the stone its name. The stone still carries faint traces of red paint suggesting the stele was more elaborately decorated. The stone now lies on one side, having been incorporated into the fabric of the later phase of Iron Age settlement, but would have stood upright as shown.

The fourth side is undecorated suggesting it was placed out of view, perhaps against a wall (Photograph: the author).

Fig 6.6 (left) The well-known pillar with incised têtes coupées. The pillar, possibly dating to the early sixth century, was discovered, like the bloc au épis above, on its side, re-used in a second-century building in Entremont. (From Arcelin and Rapin 2003, 188).

Fig 6.7 The tête-bêche motif appears as an isolated motif on this stone from Badasset (right). More often it was applied to the early engravings of human heads in rows (From Chausserie-Laprée 2000, 105).

Fig 6.8 Lintel from Nages. These later têtes coupées were engraved in heavier relief. The figures are less stylised but remain identical and anonymous. (From Arcelin and Rapin 2003, 190).

Fig 6.9 Lintel from Glanum with alternating disembodied heads and running horses. (From Arcelin and Rapin 2003, 190).

Pillars and Lintels with Cephaliform Niches

In the early-third century BC the small settlement of Roquepertuse became the site of the earliest known enclosed, indigenous ritual-space in the southern French Iron Age (Boissinot and Gantès 2000, 253). The modest curtain was enlarged, faced with masonry which included cup-marked Bronze Age stelae and fragments of Early Iron Age statuary. This new rampart was equipped with a wide double gate from which a monumental staircase led up into the sanctuary (*ibid.*, 254, 262). The site itself was levelled to form three main terraces on which worked-stone elements, dating from the Early Iron Age to the third century BC, were arranged to form a *heróon* or 'hero sanctuary'. Among the lapidary fragments found at the site were the frieze of horses' heads, the remains of several warrior statues and various sculptures including a bird of prey and the double-headed sculpture described below (Coignard and Coignard 1991; Lescure and Gantès 1991).

A central component in this sanctuary appears to have been a double portico. This monument comprised three pillars, with cephaliform niches, evidently made to hold

actual human heads. Some of these niches were smoothly carved and shallow, evidently created to house facial bones. Others were deep, with an internal projecting boss, which Coignard and Coignard (1991) suggest may have acted as a ledge on which to sit the occipital bones of a complete skull (*ibid.*, 29). Two niches still contained human facial bones, trapped in situ when the portico fell during the iconoclastic event mentioned in the previous chapter. These 'masks' were apparently coated in a clay plaster (Coignard and Coignard 1991, 27-29; Gérin-Ricard 1928, 4) and painted. A lintel, which may have acted as a cross-beam on the portico (Lescure 2004, 45), also bore cephaliform niches, and was brightly painted, with real and fantastical animals (fig 6.10). Hollows imprinted on the ground-surface of the sanctuary terrace have indicated where the pillar-sections were assembled into one monument (Boissinot 2004, 55).

The construction and assembly of the Roquepertuse portico appears to have been a collective event. The niches on each pillar were not aligned with one another, suggesting that they were created while the structure was still in pieces (Coignard and Coignard 1991, 28).

**Face A
(Front)**

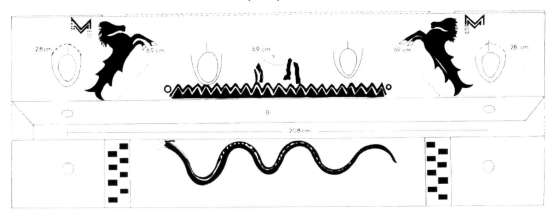

**Face B
(Underside)**

The level of workmanship in the creation of the niches also appears to vary, implying that they were the work of several individuals. Conversely, Barbet (1991) observes that the geometric patterns which decorated the monument indicate that the portico was painted after its assembly (*ibid.*, 71) (fig 6.11). Although the portico appears to have been assembled and painted on site, the mortise and tenon joints which held the structure together, also seem to have been the work of more than one person, and are badly fitted to one another. Bessac (1991) comments that it is generally considered preferable in such constructions that one person carries out this kind of work in order to ensure that the joints correspond well (*ibid.*, 50).

Pillars with niches occur on several settlement sites of the Second Iron Age (e.g Arcelin 1992; Bessac and Chausserie-Laprée 1992). Glanum in particular has produced at least 20 examples which are believed to have been in use, if not actually produced, from the beginning of the 'Hellenistic' phase of occupation, around 200 BC, to the Roman destruction in around 125 BC (Roth-Congès 2004, 29-31). All of the niche pillars from Glanum were found incorporated into monuments dating to the period of Romanisation, after 125 BC.

As at Roquepertuse, some niches were carved with an internal projection, perhaps a device designed to give added purchase to clay, cob or some other fixative (fig 6.12). At least one lintel had been fitted with an iron point or spike, presumably for the same reason. Some of the niche pillars from Glanum are simple and unadorned, however, many, like those from Roquepertuse, still bear evidence of paint and engraving (*ibid.*, 29). Arcelin and Rapin (2003, 188-89) suggest that the incorporation of human cranial elements into stone pillars may begin around the end of the First Iron Age (early- to mid-fifth century BC), as there is no evidence for pillars with head niches before then. The display of such pillars persisted well into the second century: after 150 BC at Entremont

(Salviat 1993) and up until 125 BC at Glanum (Roth-Congès 2004). The use of niche pillars as appropriated ritual furniture incorporated into later monuments seems to have continued, at least at Glanum, into the Roman period, until the final destruction of the site around 90 BC.

Fig 6.10 (top) Reconstruction of the niche-lintel from Roquepertuse. The 'waves' on which the central animal (horse?) stands are similar to the zigzag line beneath the horses heads on a lintel from the same site.

Fig 6.11 (above) The most complete pillar (pillier III, from the centre of the Roquepertuse portico). UV florescence analysis revealed red, black and white paint. The careful decoration belies the rather 'cobbled together' nature of the structure. (Both from Bessac 1991, 63).

Fig 6.12 Single niche in a pillar from Glanum where many such monuments were discovered. Internal bosses were, presumably designed to facilitate the securing of the skull.

The nature of a ritual practice, which appears to involve exposing bone to the harsh Mediterranean climate, means that it is largely left to these and other such devices to shed light on the form of cranial remains displayed in southern France during the Iron Age. The subsequent re-use of these pillars confounds the difficulty of reconstructing these practices (From Benoît 1955, 43).

Heads on Capitals

Roquepertuse and Glanum

The best-known example of this kind of architectural carving may be the double-headed sculpture from Roquepertuse which was equipped with a tenon joint for attachment to a lintel or pillar (fig 6.13). This is thought to be one of the earliest elements in the Roquepertuse assemblage, possibly dating to the fifth century BC

(Arcelin and Rapin 2003, 191; Lescure 2004, 46). Ultraviolet florescence has revealed that the sculpture originally bore painted detail (fig 6.14) (Rapin 2003, 232). Early in the site's investigation, this sculpture was designated 'Hermes' (Jacobsthal 1930) and has since been commonly called the 'Janus' figure; however, despite comparisons with Classical sculpture (Benoît 1955, 40) the characteristics of the figure appear to be indigenous (Lescure 2004, 46).

The third-century BC capitals from Glanum (Salviat 1989, 504) are more overtly syncretic. Most of these were discovered in fragments within second-century monuments, however, two were found in a more complete form within a structure of porticoes (where two skulls, discussed in the following chapter, were also discovered). The Glanum capitals had four sides, each bearing a carved head flanked with vegetal ornamentation, and topped with corbels of acanthus leaves. The heads coupled Mediterranean subject matter (satyrs, pans, Dionysus-Bacchus and the Cyclops) with indigenous expression (broadly delineated features, torcs etc.) (*ibid.*).

Heads Carved in Ronde Bosse

The depiction of human heads in *ronde bosse* (three-dimensional, plastic sculpture) is most commonly seen in the carved severed heads which sometimes accompany the third-century statues of seated warriors from Entremont, described below (Salviat 1993). Plastic sculpture of *têtes coupées*, however, continued in southern French iconography subsequent to the Roman annexation (beyond the chronological range of this study), appearing occasionally in Gallo-Roman funerary art, such as the Tarasque of Noves (a monster with two human heads beneath its claws) (Arcelin 2005, 167) and the Sphinx from Orange (crouching over a human skull) (Mignon 2005, 144).

Engravings on Architectural Masonry

In addition to the head pillars and warrior statues from Entremont, several fragments from an, as yet, unknown structure have been found on the site. A group of three decorated, architectural blocks were found on a different area from the other sculptural remains, with no archaeological context. Arcelin and Rapin (2003, 192-94) suggest that that they may date to the period of monumental building at Entremont in the second century BC. All the blocks were undecorated on one side, and bore engravings on the remaining three (fig 6.15, fig 6.16 and fig 6.17). The blocks are thought to belong to one or more columns, possibly flanking a monumental doorway, and all three are created along the same principal. The central face (A) depicts a warlike or hieratic activity: warriors on horseback on blocks 1 and 2, and a naked figure approaching a doorway on block 3. The flanking sides are carved with heads: face B depicts a single head (with open eyes on block 1 and with closed eyes surrounded by a spiral on block 3). Face C is engraved with smaller, multiple heads (Salviat 1993, 215-19).

Fig 6.13 the double-headed 'Janus' capital from Roquepertuse. (From Cunliffe 1997, pl XXI)

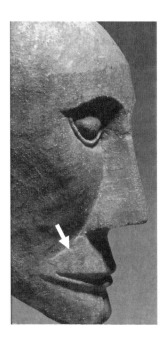

Fig 6.14 (right) Reconstruction of the double-headed capital from Roquepertuse. The capital is usually shown as plain red sandstone but traces of black paint are still visible on the sculpture. Ultraviolet florescence analysis permits a more detailed reconstruction.

The heads are surmounted by a 'leaf crown' similar to those seen in southern German sculptures such as the stone head from Heidelberg, Baden and the impressive 2.5m high, Janus-faced statue from Holzgerlingen

(From (left) cover photo, Voyage en Massalie 1990, Edisud and (right) Rapin 2003, 232).

Fig 6.15 Block I from Entremont (0.37m x 0.35m). A figure with open eyes, a warrior on horseback (an object generally supposed to be a human head suspended from the harness), and two heads with closed eyes. (From Salviat 1993, 215-17)

Fig 6.16 Block II from Entremont (0.54m x 0.34m). Much of the detail on this stone has been obliterated; however, the isolated head appears to be surrounded by a spiral similar to that encircling the figure on block III. (From Salviat 1993, 218)

Fig 6.17 Block III from Entremont (0.52m x 0.32m). Although the figure at the door has no direct equivalent in southern French iconography, there may be some parallels between this, apparently allegorical scene, and the use of porticos as ritual furniture. (From Salviat 1993, 219)

6.6 Têtes Coupées: Motifs and Themes

The early engraved *têtes coupées* on pillars and stelae appear to adhere to the same basic principle of rows of identical and schematic heads. The formulaic rendering of these heads appears to suggest an eschatological interpretation for this motif (Arcelin and Rapin 2003, 190). The heads are often seen as indicating the dead (Benoît 1964; Arcelin and Rapin 2003, 191), with closed eyes and closed or missing mouths. As such, they have been interpreted as representing war-trophies (Salviat 1993, 212) or the dead as part of a chlithonic/ancestor cult (Benoît 1955). Arcelin and Rapin (2003) have suggested that while some authors have conflated these interpretations into a nebulous 'cult of the head', the display of the heads of enemies and those of ancestors should be seen as distinct and, probably, coeval practices (*ibid.*, 189). Supplementary themes in these depictions of human heads are (frequently) the *tête-bêche* and, (on one occasion), an ear of wheat. The reversed head in the *tête-bêche* motif is often considered to carry some kind of chlithonic significance, in which the upside-down head is descending into the Otherworld (e.g. Aldhouse-Green 1999, 100). If this is the case, then the upright heads must be seen as ascending from the underworld, perhaps as part of the continuum of the cycles of life and death which Arcelin and Rapin (2003) see in the second-century architectural remains (below). The ear of wheat on the *bloc aux épis*, with its connotations of fertility, may support this interpretation, and, indeed, the dead *tête-bêche* sprouting a column of living heads, in some ways resembles a schematic ear of grain, growing from a buried seed. The horse or mounted warrior motifs also occur on pillars engraved with human heads, suggesting an early pairing of the warrior and *tête coupée* motifs. As discussed further below, the warriors often appear in funerary contexts, perhaps reinforcing the relationship between the *têtes coupées* and death.

In contrast to the early pillars and stelae, depictions of heads from the Second Iron Age appear more varied and individualistic. Those on the second-century architectural masonry at Entremont, for example, are engraved with different hairstyles and expressions (Arcelin and Rapin 2003, 192). The careful rendering of open and closed eyes on the Entremont blocks (see figs 6.15 – fig 6.17), suggests that the relationship between the living and the dead was a central theme. Arcelin and Rapin (2003) interpret the scenes on these blocks as allegorical representations of the cycle of life and death, perhaps suggesting the belief in reincarnation (*ibid.*, 190). Supplementary motifs accompanying representations or displays of heads in the Second Iron Age are also more varied. Pillars with niches seem to have been engraved and painted with a much wider range of geometric, anthropomorphic and zoomorphic figures than the earlier pillars engraved with heads. Barbet (1991) suggests that these additional images constitute a greater use of Mediterranean imagery, citing the fish-tailed horses and other mythological beasts on the Roquepertuse portico (*ibid.*, 77).

6.7 Anthropomorphic Statuary

The forerunner of the warrior representation which pervaded the Iron Age statuary of Provence can be found in the Late Bronze Age iconography of southern Europe. Around this time, anthropomorphic stelae which had once represented male and female figures became predominately male, engraved with armour and weaponry (Chenorkian 1988; Guilaine and Zammit 2001, 241-51). During the same period, rock art from the Alps to the Pyrenees, such as the large panels at Val Camonica on the north Italian Alps, and Mont Bego on the border between France and Italy, depicted weaponry and scenes of battle (Anati 1994; Harding 1994, 324; Hameau 2005, 157-59). Many of the motifs shown on these Alpine sites, in particular the images of horses and warriors on horseback, have been found on southern French pillars, stelae and lintels at Glanum and Mouriès (fig 6.9). The figures on these monuments have been simply delineated and pecked with stone tools rather than incised with metal ones, suggesting a date concurrent with the *bétyles* of the eighth – seventh centuries BC. They portray figures on horseback, engaged in hunting or combat with spears or javelins, or simply surrounded by herds of horses or deer (Arcelin and Rapin 2003, 193).

Warrior Statues of the First Iron Age

In southern France, *ronde-bosse* sculpture is first seen west of the Rhône, in eastern Languedoc, in the eighth and seventh centuries BC (Py 1990, 816-19; Arcelin and Rapin 2003, 202). These sculptures take the form of busts: near life-sized heads or, occasionally, the heads and torsos, of warriors in helmets and armour (fig 6.18). The workmanship, and the weaponry and ornaments depicted on these figures, have counterparts in contemporary central Italy (Papi 2001, 121; Arcelin and Rapin 2003, 196-202), which may reflect the extensive Etruscan activity west of the Rhône during that period. The busts appear to be more developed forms of Late Bronze Age funerary stelae, and are believed to depict 'portraits' of heroic, dead warriors, rather than deities (*ibid.*, 197). These figures continued to be created throughout the Iron Age and into the Gallo-Roman period (first century BC) (Arcelin and Rapin 2003, 203).

The earliest known statues proper (life or near-life size, in *ronde-bosse* and depicting a complete human figure) appear to be two examples, also from Languedoc. The first, from Aude (fig 6.19) was discovered, out of context at the foot of the oppidum of Le Carla in Bouriège (Barruol et al. 1961). Associated finds (Séjalon 1998, 1-2) suggested that this statue was created around the end of the First Iron Age (c. 475/50 BC). The morphology of the plinth, however, seems to indicate a slightly earlier date (Arcelin and Rapin 2003, 206), estimated to be the early-fifth or late-sixth century (*ibid.*, 212). The statue was very worn and in several fragments at the time of discovery, and the head was lost in the 1950s, however, contemporary records suggest that it was wearing a long helmet like that of the Anastasie bust and, therefore, an Early Iron Age warrior (Arcelin and Rapin 2003, 206).

Fig 6.18 (above) This warrior bust from Saint-Anastasie has stylised features redolent of the Bronze Age stelae, however, other, more naturalistic examples have been found at Nîmes and Herault. (From Arcelin and Rapin 2003, 196).

Fig 6.19 (right) Warrior statue from Aude. This early warrior statue (late-seventh/early-sixth century) holds a living, disembodied head. Although the head of the statue is now lost, the original description suggests that it wore a long helmet like that of the Saint-Anastasie bust, above. (From Arcelin and Rapin 2003, 206).

Certain features tie this sculpture to the warrior statue tradition of the Lower Rhône Valley. Its warrior persona and its stance, cross-legged on a low seat or plinth, is an enduring and common characteristic of such statues. The figure also holds a human head in its lap, which is a recurring theme in the statuary from Entremont. Nevertheless, the long flowing hair and beard, still visible on the figure's torso, distinguishes it from all other warrior representations of the period. In addition, the disembodied head appears to be alive and smiling. This appears in stark contradistinction to the majority of depictions of 'severed' heads in Provence, which, whether engraved on stelae or as part of a *ronde-bosse* composition, are often clearly intended to indicate death. Arcelin and Rapin (2003, 206) suggest that this last feature represents the head as a vessel for the soul rather than as a war-trophy (*ibid.*). The other Languedocien example is the recently excavated (2002) warrior statue from Lattes. The figure had been cut down for use as a door-post in a third-century BC house, however, the armour indicates that the figure was created during the early-fifth century if not before (Dietler and Py 2003, 790). The style of this figure, very naturalistic, near-life sized and in armour; corresponds in many respects to the warrior statues of Provence. Rather than sitting cross-legged, however, the Warrior of Lattes appears to be 'genuflecting' on one knee, in the act of drawing a bow or throwing a spear (fig 6.20) (Dietler and Py 2003). The above two statues seem to represent the sole manifestations of the warrior statue 'genre' in the Languedoc.

Fig 6.20 the Warrior of Lattes. This sixth-century warrior statue was discovered cut down for use as a door lintel in the large Languedocien settlement of Lattes. It is the only known figure of its kind. The angles of the warrior's pelvis and thighs suggest that the figure was kneeling or genuflecting on one knee rather than sitting cross-legged. The hypothetical reconstruction of the figure as an archer is based on similar indications of posture among Greek statues (From Dietler and Py 2003, 785)

Warrior Statues of the Second Iron Age

Fifth-Century Statues

Around the early- to mid-fifth century BC (the transition from the First to the Second Iron Age) sculptures similar to these Languedocien examples began to be created in the Lower Rhône Valley (Arcelin and Rapin 2003, 212) (see fig 6.1). There are around 20 known statues in this area, all in fragments, and none of them recovered in their entirety (*ibid.*, 203-205). Reconstructions of the statuary of this period are usually taken from the three most complete examples: two statues from Roquepertuse (which produced evidence for around ten warriors in all) and one from Glanum (which had evidence for three) (Arcelin and Rapin 2003, 204; Rapin 2003; Roth-Congès 2004). It would seem (from what can be ascertained of their appearance) that smaller fragments found at various sites throughout the region, such as Pierredon, Calissanne and Constantine, correspond with the main characteristics of these more complete sculptures from Roquepertuse and Glanum (Charrière 1980; Arcelin and Rapin 2003,186, 203).

The statues depict near-life sized warriors in armour, sitting cross-legged and often still bearing traces of painted decoration (Barbet 1991). The figures from Roquepertuse have also been polished and elaborately engraved and pecked to suggest armour and perhaps even skin-tone (Nerzic 1989, 14; Bessac 1991). The figures are very naturalistic, although the Roquepertuse warriors appear very slightly elongated and stylised (fig 6.21 and fig 6.22). As mentioned above, these sculptures have been deliberately destroyed, and no heads have been recovered from any statues known to have been created during the fifth century. Hands also have often been removed or obliterated (Nerzic 1989, 14) perhaps eradicating a vital clue to the symbolic or ritual function of these effigies. The extended right hand on the most complete Roquepertuse statue had been pierced vertically for the insertion of a metal accessory, while the left hand had rested on the chest (Arcelin and Rapin 2003, 205).

The best-preserved statue from Glanum (fig 6.23) is believed to date to around 500 BC, making it, perhaps, the oldest example of this tradition. The breast-plate and arm-ring all date to the end of the First Iron Age, as do architectural details on the plinth (Guillaumet and Rapin 2000; Arcelin and Rapin 2003, 204-205; Rapin 2003, 233-234; Roth-Congès 2004, 35). A counterpart to the warrior's carved torc has also been discovered in the Vix burial which dates to the beginning of the fifth century (Chaume 2001; Rapin 2003, 228). In addition, the Glanum warrior has an open tunic revealing an erect phallus, a feature which Arcelin and Rapin (2003) also interpret as an archaism (*ibid.*, 205; see also Bonenfant and Guillaumet 2002; Chaume and Reinhard 2003). There are several mid fifth-century depictions of armour from Central Europe of the same style as the armour worn by the Roquepertuse figures. The most compelling example may be the figure on the flagon from the Glauberg burial (fig 6.24), dating to around 450 BC,

which even shares the distinctive 'Buddha' posture (Rapin 2003, 228).

At both Roquepertuse and Glanum there is a continuity of several centuries between the creation and final context of these figures, but little evidence to suggest their original, fifth-century purpose, location or mode of display (Arcelin and Rapin 2003, 206). At Glanum, founded in a narrow pass through the Alpilles in the sixth century (Roth-Congès 2004), a grotto or shelf was cut into the north-eastern slope of the valley, overlooking the sacred spring which gave the site its name (Salviat 1990) and the entrance of the settlement (fig 6.25 and fig 6.26). This shelf directly opposes a natural, but apparently modified, cave in the north-western slope, which contained the remains of, at least one, worked stone block embedded in an inner corner (fig 6.27 - 6.29). Roquepertuse too has a large, apparently natural cave or recess, in the limestone stack against which the settlement was constructed (fig 6.30). The use of these grottos or caves cannot be dated, nor can they be definitely associated with sculptural remains. Nevertheless, the natural outcrops rising up to the Glanum grotto and the Roquepertuse cave have both been terraced in a manner which serves no practical purpose. Neither of these terraces creates a usable platform, and support or shoring-up of either slope is patently unnecessary, however, terraces would have enhanced these features as places of display (fig 6.26 and fig 6.31).

As regards the third- and second-century use of these statues, only Glanum has produced contextual evidence (Roth-Congès 2004, 23). The settlement underwent an episode of destruction and subsequent 'Hellenisation' around 200 BC. From this period, Glanum saw a phase of construction including a monumentalisation of the sacred spring, and the construction of various 'temples' and public buildings. In around 125 BC the settlement produced evidence of a Roman assault after which the site underwent further development. Existing structures were enlarged and new monuments were constructed, until a final Roman assault and destruction brought an end to the settlement in around 90 BC (*ibid.*, 32). The base of the oldest and most complete warrior statue at Glanum was mortised, and the fragments were discovered next to a pillar topped with a stone tenon joint (Rolland 1968, 28; Roth-Congès 2004, 26). The statue appears to have been displayed on this pillar in a warrior sanctuary established during the Hellenisation of the settlement, between 200 – 125 BC (Arcelin 1991; Roth-Congès 2004, 35-36). This *heroôn* was arranged in front of the '*porte charretière*', the monumental entrance to the settlement, which consisted of a walled passageway faced with Late Bronze Age and Early Iron Age decorated stelae (Barbet 1991; Paillet and Tréziny 2000; Garcia 2004, 112-13) until the destruction of the settlement, around 90 BC (Arcelin 1991; Roth-Congès 2004, 35-36). Fragments of another warrior statue incorporated into an edifice from the 125 BC phase of construction (Roth-Congès 2004, 35-36) suggests that, during the second century at least, the statues were not displayed at the same time.

Figs 6.21 and 6.22: Front and back views of the best preserved statue from Roquepertuse. Fragments suggesting at least another ten statues of this kind have been recovered from the site. Despite their similarities, Bessac (1991) proposes that they were the work of several sculptors. (From Gantès 1990, 165).

Fig 6.23 Warrior statue from Glanum, believed to date to around 500 BC. This figure was displayed in a herôon in the second century while two other similar figures had been incorporated into later structures. (Photograph: the author)

Fig 6.24 Warrior figure on a flagon from the fifth century Glauberg burial wearing similar armour and in a similar posture as the Roquepertuse warriors. (From Rapin 2003, 228)

Fig 6.25 (left) Artificial grotto at Glanum overlooking the Gateway and the sacred spring. Fig 6.26 (right) Terracing on the rocky slope leading up to the grotto (above) (Photographs: the author).

Fig 6.27 (above) photograph of natural cave on the northwest slope of the valley at Glanum, taken from the artificial grotto overlooking the gate of the oppidum.

Fig 6.28 (left) The cave is natural, however, a worked stone block is wedged into an inner corner suggesting that the site had, at one time been modified for some purpose. (Photographs: the author)

Fig 6.29 (right) looking from the 'natural' cave on the northwest slope of the valley of Glanum.

A = the artificial grotto,

B = the rampart (which continued to be modified and reconstructed at the same location from around the sixth century BC),

C = the spring.

(Photograph: the author)

Fig 6.30 (above left) Natural cave from Roquepertuse. The fifth century warrior statues may have been displayed in this grotto. It is difficult to resist the idea that a skull may have been displayed in the alcove above though there is, of course, no evidence for this.

Fig 6.31 (above right) Terracing on the slope leading up to the natural cave at Roquepertuse. (Photographs: the author)

Third-Century Statues

The apparent re-use of the fifth-century warriors around the third/second centuries BC (described above) was accompanied by the resumption of their creation (Arcelin and Rapin 2003, 207-10). However, once again, there is a gap of over a century between the creation of these new sculptures and the eventual context of their deposition, in the second and first centuries BC (*ibid.*, 212). This later tradition of statuary was discovered in a more localised area than that just described. All the examples come from a few sites in the immediate vicinity of Massalia, in what is usually considered Saluvian territory (fig 6.2). Petrologic analysis suggests that they were also created in this locality (Salviat 1993, 166).

The La Cloche Statue

One of the earliest examples of third-century warrior sculpture (the surviving right hand bears a sculpted ring which greatly resembles a ring from Münsingen-Rain dated to the start of the third century BC) is the figure from La Cloche (Arcelin 2004, 77-78). While the fifth-century sculptures were carved in one piece, the La Cloche warrior, like those from Entremont (below) was a composite creation, with the figure and the low seat sculpted separately (Salviat 1993, 194). Otherwise the La Cloche warrior appears to be identical in style to the iconography created two centuries earlier, even to the piercing of its hand for the insertion of some additional artefact (*ibid.*, 196). The face was damaged but the features are recognisable, through reference to numismatics, as those of Apollo (fig 6.32 and fig 6.33) (Chabot 2004, 33 and 153). The figure was discovered in the destruction level of the first-century BC oppidum, and, as with the earlier figures, there is little evidence as to where and how the statue was exhibited and/or curated during the centuries between its creation and its known display (Chabot 1983; 2004).

The Entremont Statues

The Entremont sculptures can be divided into two successive phases, again dated by reference to typologies of the weaponry, armour and jewellery which decorated the figures. The earliest group is also the largest, dating to the mid-third century BC, and comprises life-sized figures sitting cross-legged on low seats. The most complete fragments comprise five male heads and eight male torsos in armour, and three veiled female heads and two torsos in carved drapery, presumed to be female in recent analyses of the material (fig 6.34 - fig 6.38). A further 10 or 11 sculpted heads (described below) were trophies, held by the warriors (Salviat 1993; Arcelin and Rapin 2003).

The statues' heads are lifelike, though slightly elongated, and each bears enough individuality (varieties of hair styles, helmets and ornaments) and facial idiosyncrasy to suggest that they were portraits of real subjects (Armit 2006). Several statues appear to have been accompanied by sculptural representations of metal wine vessels, the two best preserved are carvings of a double-handled situla and a strainer situla (Salviat 1993, 199) though it is unknown if these were associated with female or male figures. Arcelin and Chausserie-Laprèe interpret these statues as representing 'the most important families of the aristocracy of Entremont at the time of the first siege (c. 123 BC)' (Arcelin and Rapin 2003, 208).

In addition to the 'aristocratic portraits' (though part of the same sculptural compositions) is a group of carvings manifestly representing dead, severed heads. These comprise a cluster of heads piled in a double row (fig 6.39) and a further five (or possibly six) isolated heads, all depicted with closed eyes and shut, down-turned mouths. Five are grasped from above by hands (evidently those of the aristocratic warriors) and three of these still retain fragments of the knee or leg on which they were resting (fig 6.40 and fig 6.41) (Salviat 1993, 199-208). These *têtes coupées* were also carved with realism and individuality, showing a variety of skull caps and hairstyles (Armit 2006). This ensemble is so numerous that the work of several different 'artists' has been identified; and certain stylistic and technical differences can be distinguished (Salviat 1987; Arcelin and Rapin 2003, 208).

Fig 6.32 and fig 6.33

Despite these differences, all the sculptures adhere to the same convention whereby the heads of the noble living are elongated and heads of the subjugated dead are rounded and squat (fig 6.42).

A second series of sculptures dating from the late-third or early-second century BC were also found. These comprised at least two horses and the torso of a horseman, all life-sized (fig 6.44). These have been dated by the warrior's shield and armour, and are recognisable as a cohesive group by the harder limestone (Salviat 1993, 166) and the more advanced techniques of their execution (Arcelin and Rapin 2003, 209). A group of less than life-sized statues was also discovered at Entremont depicting figures standing on plinths (fig 6.43). Again, this group is too damaged to be reconstructed, although, the depictions of footwear survive sufficiently to date this group stylistically to the second century BC (Salviat 1993, 234). Fragmented remains suggesting similar statues have also been recovered from Vachères, de Mondragon and Fox Amphoux (Rapin 2004, 21).

The third-century statue fragments from Entremont were recovered from the entrance road of the Ville Basse. This was the space dubbed 'the sacred way' by the early excavator, Fernand Benoît (1975, 229) in the later phase of settlement (150 – 90 BC) (Fig 6.45). The statues had been deliberately broken and used as construction material in the final phase of re-surfacing the entrance road to the settlement some time between the first assault in around 125 BC and the final destruction and abandonment around 90 BC. Arcelin has concluded that the statues were taken from other settlements (as yet unknown or undiscovered) and displayed in the new Ville Basse around 150 BC. He suggests that they were then destroyed during the first assault, which appears to coincide closely with the historical taking of the 'Saluvian capital' by Sextius in 123 BC (Strabo IV, 15) (Arcelin and Rapin 2002; Arcelin and Rapin 2003, 207).

Fig 6.34 (left) and fig 6.35 (right) front and side view of the torso of a seated warrior from Entremont. This new series of warrior statues were now depicted with offensive weaponry as well as armour.

Eight such torsos were discovered at Entremont and one at La Cloche. Elements of similar figures were also discovered at Mont Garou and La Courtine, however, these were very fragmentary.

(from Arcelin and Rapin 2003, 208).

Fig 6.36 (left) front and side views of one of the warrior statues (fig 6.42). Fig 6.37 (right) the warrior heads are very distinctive, with individualistic hair-styles, helmets and facial characteristics.

Fig 6.38 the third century sculptures at Entremont also include the first known ronde bosse representations of women. The female statues, too, appear to be individuals with different faces and jewellery. (From Arcelin and Rapin 2003, 208 and 209)

Fig 6.39 (above) 'Nest' of severed heads from the lap of one warrior (see reconstruction fig 6.42). (From Salviat 1993, 164). Fig 6.40 and fig 6.41 (below) severed heads held by two of the 'warrior aristocracy' from Entremont. (From Salviat 1993, 202).

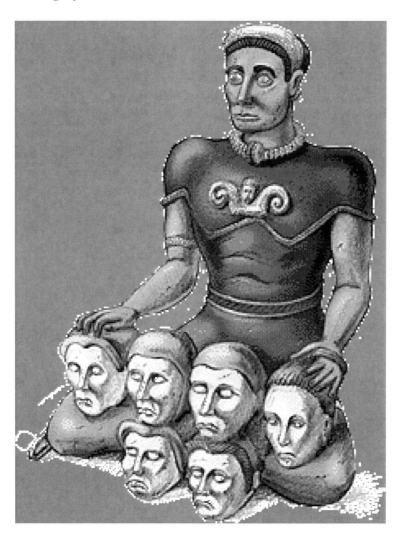

Fig 6.42 Reconstruction of one of the Entremont Warrior Statues (From http://www.entremont.culture.gouv.fr/en/f_pouvoir_her2.htm)

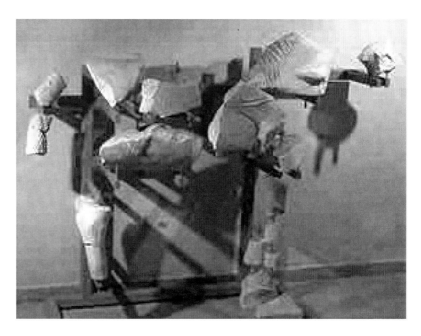

Fig 6.43 feet of a standing figure on a plinth from Entremont, dating from the second century. As discussed below, these small statues, and the figures on horseback (below) appear to represent a break from the convention of seated and particularly cross-legged anthropomorphic statuary. (From Salviat 1993, 234). Fig 6.44 the remains of at least two horses and one rider have been found, however, these are very fragmentary (From http://www.culture.gouv.fr/culture/arcnat/entremont/en/pouvoir_por.htm)

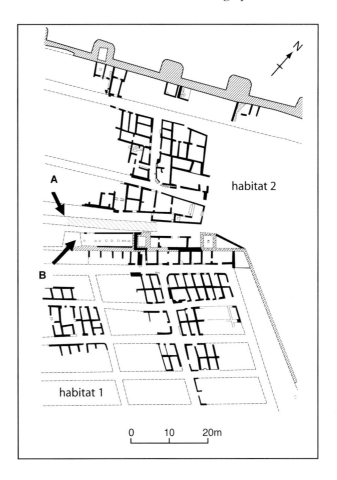

Fig 6.45 Entremont. The later foundation, 'Habitat 2', is believed to be the capital of the Saluvian confederation. The site has produced finds of iconography from the earliest Iron Age to the Roman wars.

Many of the sculptural remains were probably brought to the site during its later phases.

A The 'Sacred Way' where most of the Entremont statuary was found. The fragments had been smashed and used to metal this wide entrance way into the settlement some time between 125 and 90 BC.

B The 'Hypostyle Room', the sixth century pillar(fig 6.6) was re-used as a stylobyte and the façade may have displayed human skulls.

(After Arcelin 1990, 102)

6.8 Anthropomorphic Statuary: Motifs and Themes

The Early Iron Age

Aside from the second-century female figures from Entremont, the image of the warrior appears to have been the predominant motif of Iron Age anthropomorphic statuary in the Midi. The earliest Iron Age warrior representations in Provence are the eighth- and seventh-century engravings from Glanum and Mouriès. These images may have been symbolic, representing political power as much as military prowess, or depicting actions taking place in the Other World (Arcelin and Rapin 2003, 195). Nevertheless, they seem to embody quite literal interpretations of the warrior as a brave and spirited 'hero' (*ibid.*). By contrast, the contemporary busts found most commonly west of the Rhône appear rigid and hieratic. Arcelin and Rapin (2003, 195) suggest that the decorative details on the armour of these figures indicates high social rank. In view of certain morphological similarities to Bronze Age stelae, these busts are regarded as funerary sculpture (*ibid.*, 197). Both these early forms of iconography appear to signify the ascendancy of an image or persona of the warrior, prior to the sixth century BC, as heroic, possibly sacred, and certainly socially dominant.

The earliest identified warrior statues, the figures from Aude and Lattes, both come from the Languedoc and are believed to date to the early-fifth or late-sixth centuries BC (Barruol et al. 1961; Arcelin and Rapin 2003, 212; Dietler and Py 2003, 790). These statues are thematically dissimilar. Like the early Provençal stelae figures, the Warrior of Lattes is engaged in combat. This seemingly literal rendering of a warrior may be exceptional within the corpus of warrior statues in the Midi, although the destruction of so much of this material rules against generalisation. As far as can be determined, however, only the second-century BC horseman from Entremont appears similarly depicted in a dynamic pose. As discussed below, the cross-legged posture may indicate that the Aude warrior, like those from the Lower Rhône Valley, held some ritualistic significance. The apparently living, disembodied head, grasped by the statue would also appear to represent an esoteric abstraction, the meaning of which remains obscure.

Warrior Statues in the Fifth Century BC

Certain motifs apparent in the statues created in the fifth century BC, and in comparative, contemporary imagery (below), suggest that these were more than secular creations. The frequent use of the cross-legged posture in particular, corresponds with a convention found in other Iron Age or Gallo-Roman iconography. Figures seated in this way are depicted on various ritualistic or ceremonial objects in temperate Europe (Guillaumet 2003) such as the horned deity on the Gundestrup cauldron (Taylor

1992), the figure on the Glauberg flagon (Herrmann 1997) and the votive statuette from Bouray-Sur-Juine (Lantier 1934). Other themes particular to the fifth-century group of statues may have similar parallels with ritualistic iconography, both in temperate Europe and the Midi. In addition to the figure on the Glauberg flagon for example, engravings of warriors similar to the statues from Roquepertuse appear on a plaque from Cavaillon (fig 6.46) depicting libation rituals. There may also be parallels between the erect phallus of the Glanum statue and near-contemporary funerary statuary, such as the Hirschlanden warrior (Megaw 1970, no. 12).

Rapin (2003) proposes that the entire corpus of Provençal warrior statuary was a form of funerary iconography (*ibid.*, 244). The piercing of the statues' hands (Arcelin and Rapin 2003, 205) could indicate that, rather than acting as simple *memento mori* or burial markers, the statues may have been used in ceremonies, perhaps involving the insertion and removal of artefacts at significant times. Parallels with other forms of iconography appear to suggest that the fifth-century statues in particular, expressed themes of death and burial. Wine vessels, in libation rituals (depicted on the Cavaillon plaque), or as offerings to the dead or a chlithonic deity (as is presumably, is the case with the Glauberg flagon), may also have had some relationship to in the figures from Roquepertuse. As regards any more literally warlike significance in these representations of warriors, it has been noted that both the Glanum and Roquepertuse figures wear armour, but carry no offensive weapons (Arcelin and Rapin 2003, 205).

Warrior Statues in the Third Century BC

The statues from La Cloche and Entremont clearly arise from the same tradition as those created in the fifth century BC. Both groups follow the same conventions of form, including the detail of the pierced hands, suggesting that, perhaps, they functioned in a similar way. However, differences in technique, most notably the use of composite carving at La Cloche and Entremont (Salviat 1993, 194) and the lack of evidence for the creation of such figures during the fourth century (Arcelin and Rapin 2003, 212), imply a different phase of construction, and perhaps, a different significance.

The Entremont assemblage represents the largest collection of anthropomorphic statuary in the Midi. Many of these third-century statues have motifs in common with the early sculptures, notably the inclusion of human heads and wine vessels. Unlike the statue from Aude the severed heads in the Entremont compositions are clearly dead, and in attitudes of subjugation. The Entremont heads also differ from the *têtes coupées* on the early *bétyles* and pillars, in that they are not identical and anonymous. The Entremont heads appear distinctive enough to imply that they represented either specific individuals, or discrete tribal or ethnic groups (Armit pers comm).

The earliest and most concentrated representations of female anthropomorphic statuary were also found at Entremont. There is no evidence that these statues represent either deities or priestess, having no other attributions than displays of wealth through decorative head-wear and jewellery. This conclusion concurs with Arcelin and Rapin's interpretation, that these figures represent Late Iron Age aristocracy (Arcelin and Rapin 2003, 208). This may signify a move away from the representation of, presumably, powerful figures, exclusively through the persona of 'the warrior'. It also appears to indicate that by the third century, indigenous power was sometimes expressed through overtly political, rather than ritualistic, themes.

The large, sculpted jars which accompanied some of the Entremont figures may indicate a continuity of the libation and ritual-offering themes suggested in the earlier iconography. Alternatively, these vessels may represent the accoutrements of the feast, and, therefore, a display of status and hierarchy; or symbols of the agricultural produce from which this aristocracy derived its wealth. These interpretations are not all necessarily mutually exclusive, although, in the absence of other supporting evidence, they are, of course, largely speculative.

As mentioned above, the third- or second-century figures on horseback from Entremont appear to signify a revival of representations of the warrior as a combatant. Rapin (2004) proposes that these horsemen and the finds of sculpted feet standing on plinths which may also date to around this period, indicate a deliberate and conscious break from the tradition of representing 'ancestors' in a posture associated with divinity (*ibid.*, 21). This suggests a continuation of the 'politicisation' of the ruling classes already, apparently, underway by the third century BC.

6.9 Ritualistic Sculpture and Engravings

Various sculptured and engraved stone fragments discovered at settlements in the Lower Rhône Valley cannot be said to belong to any one cohesive, lapidary tradition, consisting as they often do, of stray or re-used remnants from vanished structures. The lack of comparative material means that dates suggested for these pieces are sometimes more speculative than for the anthropomorphic sculpture and depictions of heads above.

Ritual Practices

Representations of ritual practices rarely occur in southern French Iron Age iconography. Of the two known examples, the frieze from Cavaillon is the earliest and most securely dated. This stone plaque depicts a libation ritual, comprising a procession of figures carrying large, two-handled vessels (Arcelin and Rapin 2003, 194). The frieze has no real parallels within indigenous iconography, and has been likened to scenes from Early Iron Age Italian situla-art (Arcelin and Rapin 2003, 194; Lucke and Frey 1962, 59). The central image was bordered by incised and painted geometric patterns corresponding closely to the motifs decorating the armour of the fifth-century BC warrior statues of Roquepertuse (Barbet 1991, 54-55) (fig 6.46). The second example of iconography depicting ritual activity is the fragment of a lintel which was recovered from a second-century horizon at Entremont. The original date of its creation remains unknown, however, the stone is engraved with the same naturalistic qualities as other elements dated to the second century BC (Arcelin and Rapin 2003, 194). The lintel depicts two figures which appear to be female. They are carrying small animals in their right arms, and their left arms are raised in a gesture associated with sacrificial offering (Salviat 1993, 225; Arcelin and Rapin 2003, 194) (Fig 6.47).

Fig 6.46 (right) Frieze from Cavaillon. Despite comparisons to Italic iconography, this frieze may reflect the public form of libation rituals frequently practiced in the domestic sphere in the Midi (From Arcelin and Rapin 2003, 194).

Fig 6.47 Lintel from Entremont, perhaps from a second-century structure, which appears to show the ritual sacrifice of two small animals. (From Arcelin and Rapin 2003, 194).

Ritual Motifs

Horses appear frequently in the southern French iconography, depicted on stelae, on architectural elements or decorating lintels with engraved heads or cephaliform niches. An unusual variant of this motif is the stone lintel engraved with four horses' heads, recovered from Roquepertuse (fig 6.48 and fig 6.49). This lintel carries a tenon joint on its right-hand edge, indicating that it was once part of a longer frieze (Boissinot 2004, 55). Ultraviolet fluorescence has revealed traces of blue-green and red pigments, showing that, like the pillars and statues from the site, the frieze was brightly painted (Barbet 1991, 78). The lintel is believed to come from the

period of the sanctuary (early-third century) (*ibid.*) although it may have been an earlier, curated piece. The horses appear identical and schematic in contrast to other, more naturalistic representations such as the third- and second-century iconography from Entremont.

The sculpture of a bird, also from Roquepertuse, was found in 25 pieces scattered over several square meters (Gérin-Ricard 1927, 21). Originally thought by some to be a goose (e.g. Ross 1986, 145), the end of the beak has recently been recovered and it is now recognisably a bird of prey (Boissinot 2004, 54) (Fig 6.50). Scientific analysis has failed to produce evidence that the sculpture was originally painted (Barbet 1991, 65). Benoît (1955, 19) believed that the bird had originally sat on top of the

Fig 6.48 (above) part of a lintel from Roquepertuse (Coignard & Coignard 1991)

Fig 649 the lintel was subjected to UV florescence analysis by Alix Barbet (1991) the dark shading represents areas painted red, the lighter area were a blue-green pigment unusual in this corpus of iconography which usually bore red, white and black paint (from Barbet 1991, 63)

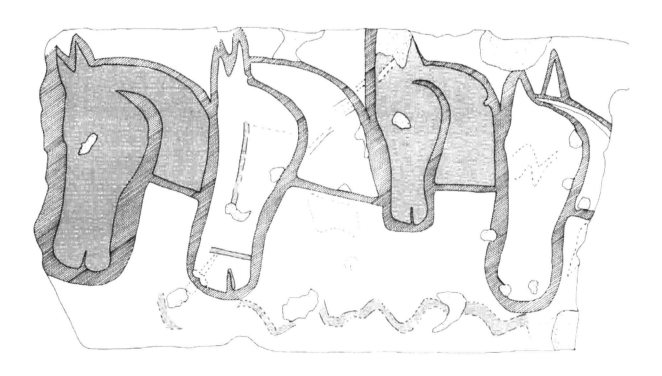

Fig 6.50 Sculpture of a Bird from Roquepertuse. (From Py 1993, 245).

portico, however, this was speculation. The sculpture was found in the level of the granary or farm which post-dated the oppidum, although, as Boissinot states, there is nothing to suggest that it was not already in pieces at the beginning of this phase (2004, 54).

Other, more uncommon motifs occur at sites throughout the Lower Rhône Valley. The ear of wheat engraved and painted on the *bloc aux épis* (described above) at Entremont appears to be unique, while the snake engraved on a pillar at the same site (Fig 6.51) has parallels at Glanum and Roquepertuse (Barbet 1991, 79). All of these elements were re-used within second-century constructions, but they appear stylistically more akin to sixth- and fifth-century iconographic traditions.

The most singular suite of motifs may be those painted on a lintel with cephaliform niches at Roquepertuse, which depicts real and mythological creatures. Boissinot proposes that these motifs are thematically closer to north Italian than Celtic or Greek influences (*ibid.*, 55). The large number of curated elements at Roquepertuse makes dating problematic, however, the association of these paintings with the sanctuary suggests that the motifs, if not the lintel itself, date to around the beginning of the third century BC (fig 6.52).

6.10 Ritualistic Sculpture and Engraving: Motifs and Themes

Ritual Practice

The Cavaillon plaque and the lintel from Entremont may show the public forms of rituals apparently practised with some frequency in the domestic sphere. Throughout the Midi, and, perhaps particularly around Martigues, the necks of amphorae, have been found, partially sunk into house-floors (fig 6.53). These appear to have acted as conduits for domestic libations (Chausserie-Laprée 2005, 231). Similarly, the figures apparently sacrificing small animals from second-century Entremont evoke the numerous finds of small animals beneath house floors throughout southern France. This practice appears

especially widespread in the Languedoc, where animals, particularly snakes, birds and rodents, were often buried under floors (Chausserie-Laprée 2005, 130-31). In the Lower Rhône Valley, this practice is most commonly seen in Martigues, but small animals have also been found in domestic deposits at Roquepertuse and other sites around Massalia (Boissinot 2004, 52). Iconography depicting ritual activity is highly unusual, and such things are not pictured on other media. Indigenous pottery for example, mostly copied Greek styles, without any distinctively native decoration; and most other iconography, the statues and porticoes described below, appear to portray posed and static symbolism.

Ritual Motifs

The horse, either as a solitary motif or as part of a composition depicting warriors on horseback, is a staple motif not only in the Midi but throughout Iron Age Europe. The connotations of elite status and military strength are obvious, and Arcelin and Rapin (2003) suggest that the horse constitutes the symbol of a military aristocracy (*ibid.*, 212). That some of the other animal motifs carried specific ritual significance is indicated by archaeological evidence for the interment of snakes, birds and other small animals, again within the sphere of domestic ritual practice (Dedet and Schwaller 1992). Barbet (1991) suggests that the birds depicted at Roquepertuse may illustrate the supposed Gaulish practice described by Strabo (IV, 4, 6) of divining the future through observing the flight of birds (*ibid.*, 78). In the context of a sanctuary dedicated to the display of human remains, however, it may be more straightforward to link depictions of flesh-eating birds to rites of exposure and excarnation.

As with the horse and bird motifs, though, perhaps to a lesser degree, depictions of serpents occur throughout Iron Age Europe. Barbet (1991) suggests that the snake bore some chlithonic significance, citing the appearance of the serpent held by the cross-legged divinity on the Gundestrup cauldron (*ibid.*, 78). This interpretation is well illustrated by the inclusion of a serpent in the 'drowning' scene on this vessel (fig 6.55), which is convincingly argued by Miranda Green (2001) to represent an allegory of death and reincarnation (*ibid.*, 114). In view of this interpretation, it is interesting that the painting of a snake from Roquepertuse appears on the underside of the lintel (fig 6.52).

6.11 Patterns of Violence and Warfare from the Iconographic Evidence

As outlined at the beginning of this chapter, the iconography of the southern French Iron Age (particularly around the Lower Rhône Valley) is dominated by images of warriors and disembodied human heads. These motifs demonstrated great longevity. Early iconography was curated and re-used, and new images of warriors and severed heads were created throughout the period. While the main conventions remained the same (cross-legged warriors and heads in rows) each re-

Fig 6.51 the carving of a serpent on a pillar from Entremont is much worn and only just discernable. Like the pillar engraved with têtes coupées it has been re-used as a threshold stone in the second century BC 'hypostyle room'. The area immediately in front is the Sacred Way where the statuary fragments were found (photo the author)

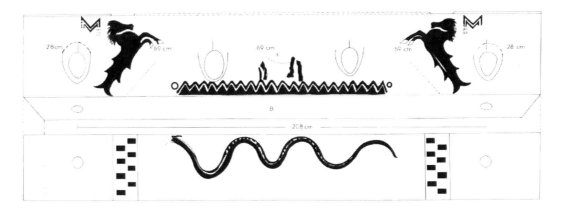

Fig 6.52 Ultraviolet florescent analysis has allowed a partial reconstruction of the lintel with cephaliform niches from Roquepertuse.

A = front face of the lintel, B = underside. Mortise joints for attachment to pillars are shown on face B. (From Barbet 1991, 63)

working of these motifs evinced subtle changes in theme which can be observed in the images themselves and in comparable iconography in Mediterranean and temperate Europe.

The tables (table 6.1 and table 6.2) summarise the main themes associated with the representation and display of warriors and severed heads. Though, strictly speaking beyond the scope of this analysis, the eighth- and seventh-century representations are included, as above, to provide a context for the emergence of the warrior statue tradition. This summary is divided into two periods: before and after the fourth century BC, to take account of the apparent hiatus in the known creation of this iconography. Admittedly, this is likely to be an artificial lacuna, resultant on differential survival and limited investigation; nevertheless, as can be seen, the two

periods also appear to encompass very distinct traditions.

Throughout the Iron Age the character of the warrior retained a sense of being emblematic of power. The manner or emphasis of this power, however, whether militaristic, ritualistic or political, changes during the period. The literally martial representations of the stelae engravings and the Warrior of Lattes, and the strongly eschatological appearance of the Aude figure had already been conceived prior to the creation of the first tradition of warrior statues in the Lower Rhône Valley. The ritualistic posture and unarmed status of these fifth-century figures appear to have continued throughout the third and (presumably) fourth centuries BC. In the third and second centuries these themes changed, initially, in the addition of trophy heads and weaponry, and ultimately, in the rejection of the ritual posture and the resumption of literal attitudes of combat.

Têtes coupées also underwent changes in theme during the Iron Age. The early pillars and stelae, like the early warrior statues, are redolent of ritualistic symbolism. The mid-fifth century may have seen the first use of cranial remains and (possibly) the appropriation of the 'heads in rows' convention by an ancestor cult. The sculpted heads from Entremont and cranial remains from Roquepertuse appear roughly coeval, perhaps supporting the supposition that distinct forms of head-ritual co-existed (Arcelin and Rapin 2003, 190). By the late-second century, pillars with niches were used as appropriated ritual furniture in later structures, however, there is no evidence that they retained their original function. Examination of the osteological remains in the following chapter may suggest that, while cranial elements continued to be curated and displayed, pillars with cephaliform niches were no longer used in this way.

Fig 6.54 Decapitated lamb placed beneath house-floor in Saint-Pierre, Martigues. Animals or parts of animals have been found deposited in pits in house floors in many sites. The practice is most frequently seen in the western part of the region, in the Martigues area of the Lower Rhône Valley and on the other side of the Rhône in the Languedoc.

(Fig 6.53 & 6.54 from Chausserie-Laprée 2005, 231)

A changing pattern of aggression may be seen in these fluctuations between iconography expressing overtly aggressive themes or iconography with more ritualistic appearance and associations. This pattern, as regards both warrior representations and the representation and display of severed heads, is compared with the patterns of aggression compiled from the documentary evidence, (table 6.3), and from the settlement record (table 6.4a and 6.4b) below. This comparison utilises only the iconography from the main study area, Provence.

The sixth and fifth centuries BC (the period of warriors and severed heads as ritualistic motifs) are poorly served by the documentary evidence. The emergence of the persona of the warrior as a military and political leader, and the depiction of heads as both trophies of subjugation and venerated ancestors, pre-date the Roman intervention by around 150 years. Interestingly, the more literal representations of warriors as combatants (seen in the eighth and seventh centuries) appear to be revived around the time of the Roman campaigns.

Fig 6.53 Collars of Italic amphorae in a second-century BC house-floor at L'Île. Chausserie-Laprée interprets these as 'receptacles or conduits for libation offerings'.

Fig 6.55 The 'drowning scene' on the Gundestrup cauldron. In Europe the serpent (depicted top right) is often associated with chthonic themes. Green argues that this panel represents scenes of death and reincarnation. (From Green 2001, 114)

8th / 7th Century	**Iconography:** (Languedoc) Armed warriors on horseback on stelae and funerary busts **Associated motifs:** Horses, deer/stags **Themes:** Warrior = combatant, military power, hunting, wealth, death
6th Century	**Iconography:** (Languedoc) Warrior Statues ritual (Aude) and warlike (Lattes) (Provence) Schematic Engraved heads on Pillars **Associated motifs:** Tête-bêche, libation rituals, wheat, horses, snakes **Themes:** Warrior = (Languedoc) military power / ritual power , death / re-birth Têtes Coupées = (Aude, Provence) chlithonic / fertility, death / re-birth
5th Century	**Iconography:** Seated Unarmed Warrior Statues Human crania in pillars? **Associated motifs:** Libation rituals, horses, snakes, birds **Themes:** Warrior = ritual power, libation rituals, death Têtes Coupées = ancestor veneration?

Table 6.1 Iconographic Evidence Prior to the Fourth Century BC.

The warrior representations during this period are overtly ritualistic, and themes of death and rebirth recur frequently. There are also similarities between the fifth century statues and iconography (often buried, as grave goods or votive offerings) connected with funerary and drinking/libation-rituals.

Between the sixth and fifth centuries, the severed head also appears to be related to concepts of life and death. Associated motifs (serpents, wheat and têtes-bêche), and, perhaps, the morphology of the heads themselves (formed like ears of wheat) suggest preoccupations with chlithonic and fertility rites.

3rd Century	**Iconography:** 5th Century Seated Unarmed Warrior Statues 3rd Century Seated Armed Warrior Statues with heads Human heads in Pillars **Associated motifs:** Libation rituals, horses, snakes **Themes:** Warrior = military strength, political power, wealth, dominance Têtes Coupées = (e.g. Roquepertuse) ancestor veneration
2nd Century **(>125 BC)**	**Iconography:** 5th Century Unarmed Warrior Statues (Glanum) 3rd Century Armed Warrior Statues with heads (Entremont) 2nd Century Warriors on Horseback / Standing Figures **Associated motifs:** Pillars with engraved heads and niches used in public buildings **Themes:** Warrior = combatant, military power, political power, wealth Têtes Coupées = (Entremont) trophy-heads
2nd - 1st Century **(125 – 90 BC)**	**Iconography:** 2nd Century Warriors on Horseback, Standing Figures (Entremont) 5th Century Unarmed Warrior Statues (Glanum) **Associated motifs:** 3rd Century Armed Warrior Statues used in roadway (Entremont) Pillars with engraved heads and niches used in public buildings **Themes:** Warrior = military power, heroism, wealth Têtes Coupées = trophy-heads (Entremont)

Table 6.2 Iconography after the Fourth Century BC.

Third century warriors were depicted with offensive weaponry and the sculpted heads at Entremont suggest themes of subjugation. Conversely, the roughly coeval head-portico at Roquepertuse may be related to ancestor veneration.

By the second century, warriors were depicted engaged in warlike activities. Iconography related to head-ritual or display (pillars with engraved heads and niches) during this phase appears to have been reused in public buildings.

Century	Documentary Evidence	Iconographic Evidence
6th Century		Head = symbol of death/rebirth
5th Century		Warrior = Ritual Leader Head = symbol of death/rebirth
4th Century	390 BC: Gallic Siege of Massalia	Warrior = Ritual Leader Heads = ancestors
3rd Century	219–02 BC Hannibal marches through Languedoc	Warrior = Military/ Political Leader Head = trophy Head = ancestors
Early 2nd Century	154 BC Q. Optimus's campaign	Warrior = Military/Political Leader Warrior = Combatant/ Political Leader Head = trophy
Late 2nd Century	125 BC M. F. Flaccus' campaign 123 BC Roman garrison at Aix 106 BC Caepio's Campaign	Warrior = Combatant/ Political Leader
1st Century	49 BC Roman siege of Massalia	

Table 6.3 Comparison between Iconographic and Documentary Evidence for Warfare.

The designation 'combatant' refers to warriors engaged in warlike activities, 'military' leaders refer to armed, seated figures and 'ritual' leaders refer to figures in armour but with no associated, aggressive motifs. The term 'political' leader refers to individualistic figures, or 'portraits' associated with motifs of subjugation.

6th – early 4th Centuries	Mid 4th – 3rd Centuries	Early 2nd Century	Late 2nd Century
Some direct evidence for aggression	Least direct evidence for aggression	Greatest direct evidence for aggression	
Some intention to defend settlements	Least intention to defend settlements	Greatest intention to defend settlements	
Larger households	Smallest households	Small households	Largest households
Some communality	Some communality	No communality	Greatest communality

Table 6.4a Simplified pattern of warfare apparent from direct and indirect evidence for aggression, and from the most impelling results for recognising fear and socialisation for mistrust.

Century	Settlement Evidence	Iconographic Evidence
6th	Some direct and indirect evidence Some communality Little fragmentation	Head = symbol of death/rebirth
5th	Some direct and indirect evidence Some communality Little fragmentation	Warrior = Ritual Leader Head = symbol of death/rebirth
4th	Little direct and indirect evidence Great communality Greatest fragmentation	Warrior = Ritual Leader Heads = ancestors
3rd	Little direct and indirect evidence Great communality Greatest fragmentation	Warrior = Military/ Political Leader Head = trophy Head = ancestors
Early 2nd	Greatest direct and indirect evidence Great fragmentation Least communality	Warrior = Military/Political Leader Warrior = Combatant/ Political Leader Head = trophy

Table 6.4b Comparison between patterns of warfare compiled from the forms of settlement evidence and the changing themes of militaristic, political and ritual power in the iconography of warriors and severed heads

6.12 Conclusion

The Warrior

The iconographic record appears ritualistic rather than military in themes relating to both warriors and severed heads between the sixth and fourth centuries BC. The warrior statues of the Lower Rhône Valley do not appear to reflect either the vague historical references to conflict between Gaulish tribes and the Phocaean colony, or the sporadic violence apparent in the settlement evidence around Martigues during this period. The emergence of warrior statues displaying military and political characteristics in the early-third century BC, along with the representations of heads as trophies of war (as well as revered ancestors), appears to pre-empt the settlement-derived evidence for a concentrated period of conflict around 200 BC by up to a century. Only the ultimate re-invention of the warrior statue in the second century BC, which depicts the warrior as a literal combatant, appears to broadly coincide with the period of Roman aggression in the region, in the second half of the second century BC.

It is possible that the appearance of warrior statues displaying more martial themes around the beginning of the third century indicates small-scale internal warfare in the Lower Rhône Valley which other archaeological or historical evidence has not yet identified. These figures may represent a situation of increasing tension prior to the concentrated outbreak of conflict revealed in the settlement evidence around 200 BC (table 5.1). This and other interpretations of the third-century warrior statues will be discussed further in chapter eight. For the present it is sufficient to conclude that an iconography of warriors does not necessarily indicate a warlike society or environment. In the case of the warrior statues of southern France, the warrior appears to represent power, but whether ritual, military or political power, requires careful analysis of associated imagery and reference to other forms of evidence for warfare.

Representations of Human Heads

As stated above, depictions of human heads may constitute the premier motif of the southern French Iron Age. As with the changing images of the warrior, the iconography of the human head must be considered in relation to its morphology, associated iconography and motifs, and the context of its display. Within a more contextual analysis, the iconography related to the human head appears to demonstrate a preoccupation with chlithonic and fertility beliefs during the Early Iron Age, and a possible appropriation of the conventions of this belief, to the service of an ancestor ritual from around 475/50 BC. The iconography itself yields no evidence that the heads it represents were heads taken in war. Arcelin and Rapin (2003) suggest that the Early Iron Age pillars engraved with heads are unlikely to have been substitute war-trophies. They argue that warriors would have considered such carvings to be poor replacements

for the flesh-and-blood trophies of the battlefield (*ibid.*, 189). As discussed in chapter eight, such 'surrogate' heads are not unknown in the ethnographic record (e.g. Hoskins 1996, 30; George 1996, 50-89). Nevertheless, the engraved rows of heads of the earlier Iron Age appear to hold esoteric rather than overtly aggressive connotations.

Heads engraved on pillars and stelae, and the later pillars and lintels designed to hold actual human crania, have prompted dissent between those who see this iconography as representing war trophies (e.g. Salviat 1993) or ancestor veneration (e.g. Arcelin and Rapin 2003). In Anglophone archaeology also, there are differences of opinion. The so-called 'cult of the severed head' has been viewed as a straightforward practice designed to denigrate the defeated enemy (Collis 2003, 216), and as an attempt to control the power of the fallen adversary (Cunliffe 1997, 210). Aldhouse-Green notes that classical accounts of Celtic headhunting suggest that heads were taken for ritual purposes but obtained from enemies killed in battle (2001, 96). She lists various possible interpretations of head-related practices among the Celts but warns against 'uni-interpretational' approaches (*ibid.*, 101). Aldhouse-Green's reading of the documentary evidence for head rituals (2001, 96), isolates a major problem in attempting to interpret this iconography as evidence of warfare, which is the frequent ritualisation of predatory headhunting. This question is explored further in the following chapters. For now, it is enough to state that the iconography related to representations of 'severed' heads in southern France cannot be accepted as proof-positive of war-related headhunting without the support of other evidence.

7.1 Introduction

Osteological evidence for violence or warfare usually comprises direct indications of violence, such as trauma and embedded projectiles in human remains, or indirect indications, in the form of mass-graves or other apparently disrespectful depositions of human remains. Funerary evidence, including such 'rhetorical' (Armit *et al.* 2006) indications of warfare as deposits of offensive and defensive weaponry, may often be seen as signifying warfare also. Essentially these forms of evidence depend on the contrast between conspicuous burials and normative practices. In southern France, only the Iron Age cemetery at Maihac (Janin *et al.* 2002), in Languedoc, has allowed the analysis of osteological trauma or the appearance of weaponry as grave goods. In this case, it has been suggested that, as weaponry was confined to adult male burials, the arms may have been operational rather than status-markers or heirlooms (*ibid.*).

Elsewhere in Mediterranean France, the opportunities for such studies are lacking, due to the current dearth of known or excavated material. In the Var region, a large, underground cave, Aven Plérimond, was used as a depository for human remains. Around 17 skulls and the post-cranial remains of several individuals have been recovered; as well as items of jewellery, armour and horse-gear; dating from the Chalcolithic to the Gallo-Roman period (Perrot 1971). Stratigraphic dating of the human remains was not possible, however; and, at the time of writing, none of the bone has been submitted for scientific dating. In addition, some, few, isolated tumuli believed to date to around the eighth century have been noted, though rarely explored, on the slopes of the Alpilles and the Sainte-Victoire (Walsh and Mocci 2003, 47).

The Lower Rhône Valley meanwhile has, as yet, yielded no examples of cemeteries or other 'normative' burials and very few mortuary deposits in the indigenous milieu, from the beginning of the Iron Age until the Gallo-Roman period.

In the Lower Rhône Valley, almost all the known finds of human remains have come from non-funerary, settlement or ritual contexts. Most notable among these are skulls and cranial elements (fig 7.1), such as those found at Entremont, Roquepertuse or Glanum, traditionally associated with the Celtic 'cult of the severed head' (Ross 1989, 121-23). In addition to these, there are occasional discoveries of infant burials beneath floors (e.g. Nin 1999), and other, perhaps more anomalous, finds, like those at L'Île and Saint Pierre (Chausserie-Laprée 2005, 227) described below.

This chapter begins with a synthesis of the human bone recovered from sites in the Lower Rhône Valley; as described by the original excavators, in osteological analyses, and, where it has been possible, from personal observation. In light of the preponderance of cranial material, and the large body of iconography suggesting that the human head bore a special significance throughout the Iron Age, the material is presented in term of cranial and post-cranial remains, however, the analysis considers all forms of osteological evidence in tandem. Direct evidence for violence, as well as the age and sex of the individuals (in reference to the recognition of violent death or the identification of warriors) will be noted; as will the chronological context and the mode of display or deposition. This chapter considers whether, in view of these factors, and in relation to the patterns of warfare compiled in the previous chapters, any of the human remains described below, can be considered a product or expression of warfare in the region.

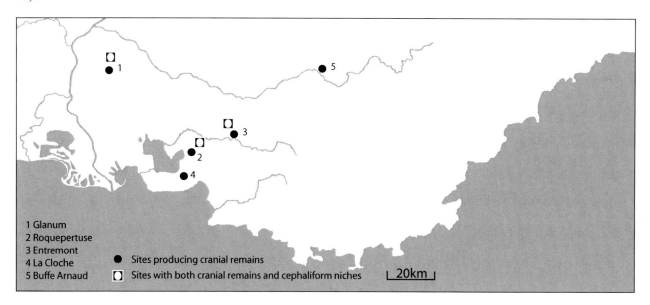

1 Glanum
2 Roquepertuse
3 Entremont
4 La Cloche
5 Buffe Arnaud
● Sites producing cranial remains
▢ Sites with both cranial remains and cephaliform niches
⌐ 20km ⌐

Fig 7.1 The Roquepertuse remains have been mislaid and the others tend towards the latter part of the period though this most likely reflects survival rates rather than the full geographical and temporal range of the practice. Three of the five sites have also produced niche pillars suggesting a longer-lived practice of head-display.

7.2 Synthesis of Human Remains from the Lower Rhône Valley

7.3 Skull Deposition and Display

Finds of complete or near-complete skulls with no evidence of associated post-cranial bone on indigenous settlement or ritual sites in the Midi are usually designated evidence for the headhunting practices of the Gauls described by Poseidonius and Polybius (Garcia and Bernard 1995, 122), and it is these remains which are described in this section.

Roquepertuse c. 300 – 200 BC

The earliest known context to produce the remains of apparently displayed or curated skulls is that of the early third-century BC sanctuary at Roquepertuse, described in the previous chapter. The bones and the context of their discovery were described by the excavator H. de Gèrin-Ricard as consisting of facial bones, remodelled in clay and painted (Coignard and Coignard 1991, 29) (fig 7.2). One of these 'masks' was found beneath a pillar of the collapsed portico, still lodged within its niche; while another two were found nearby (Gèrin-Ricard 1928, 4; Coignard and Coignard 1991, 27-29).

Fig 7.2 Neolithic plastered skulls from Jericho are often cited as comparanda for the plastered facial bones from Roquepertuse described by Gèrin-Ricard (From Bonogofsky 2004, 118).

In the mid-1950s, R. P. Charles (who carried out the initial analysis of the Entremont skulls, below) identified these remains as two males, aged between 35 and 45 years, and one individual of undetermined sex (though with feminine characteristics), which he estimated to be around 30 years at time of death. In 1986 and 1987, the discovery of new skeletal remains at Entremont prompted a re-examination of the early finds (described below). In the course of this research, a box marked 'Roquepertuse' was found, containing very fragmented elements of human crania, from which nothing could be ascertained beyond that it comprised several individuals, one of which was a male of more than fifty years (Mahieu 1998, 65). In addition to these, a calotte (skull-cap) was discovered in the vicinity of the entrance during the campaign of excavations between 1994 and 1999. No analysis has, as yet, been published on this later find (Boissinot and Gantès 2000, 267-68).

Buffe Arnaud 225 – 125 BC

The oppidum of Buffe Arnaud, founded around 225 BC, has undergone only limited investigation. To date, little is known regarding the settlement area, and detailed excavation has been concentrated around the *'tour-porche'*; a monumental, gate-tower of a design otherwise unknown in Provence (Garcia and Bernard 1995). A destruction level, dating to around 125 BC, and comprising numerous catapult boulet and widespread burning, appears to mark the abandonment of the site. Seventy-six burnt cranial fragments were discovered in the entrance, in a level directly below the destruction layer, prompting Garcia and Bernard to suggest that they were displayed on a central post in the middle of the gateway prior to the assault on the oppidum (*ibid*, 121-22). The bones comprise the cranial and facial bones of at least two individuals, one was male, over 30 years old, the other, adolescent, of indeterminable sex. The majority of the remains came from the upper parts of the crania, with fewer fragments from the bases of the skulls or the lower parts of the occipital bones. The teeth were believed to have fallen out post mortem, and no mandible-fragments were recovered. The burning on these crania appeared to be consistent with bone which had already fallen into the entrance and been covered with a thin layer of earth prior to the burning of the gate (Dudet 1995).

Entremont 175 – 90 BC

Displayed or curated cranial remains from Entremont were discovered during Benoît's excavations in the 1950s, either in the hypostyle room (Benoît's *Salle de Crânes*) or along its facade on the wide entrance-road of the Ville Basse (the *Voie Sacre*) (see fig 6.45) (e.g. Benoît 1975), both of which date to after the first Roman destruction in around 125 BC. Mahieu describes these remains as very degraded, consisting, for the most part, of fragments, stored simply by lot, and a few near-complete calva, six of which bore traces of piercing or still held iron carpentry nails (fig 7.3 and fig 7.4). The assemblage comprised an estimated minimum number of individuals of around 20. Of those which yielded an estimated age of death, the youngest appears to have been 20 – 30 years, five individuals were 30 – 40 years, five between 40 – 50 years and one man was estimated to have been over 50 years old at the time of his death. Of the six examples which were sufficiently complete to distinguish sex, all were male; however, Mahieu (1998) declines to dismiss the possibility that some of the assemblage were female.

Fig 7.3 Skull 1 from Entremont

Fig 7.4 Skull 4 from Entremont (both from Mahieu 1998, 63)

Glanum 125 – 90 BC

Glanum has produced two caches of complete or near-complete skulls from beneath monumental buildings post-dating the first Roman assault on the settlement in around 125 BC. The earliest found deposition of eight

skulls was discovered during the excavation of the 'Gemines Temple' in 1958. A second discovery of two skulls, also almost complete, was uncovered in 1967 on a level consistent with the floor of a chamber in the rear of a large edifice which appears to have had an open façade of pillars or porticos (Roth-Congès 2004, 28) (Fig 7.5). These crania have never been subjected to formal osteological analysis, indeed, three of the skulls from this assemblage, located in a box at the Musée de Glanum (fig 7.6), still contain the earth from their excavation (Ian Armit pers. comm.). Nevertheless, the excavator, Rolland, has left a serviceable description of the remains. The eight skulls were described as 'young' adults (determined from dentition and unfused sutures) with no vertebrae but some mandibles still attached (Rolland 1958, 12). The other two skulls both bore unhealed trepanation marks (fig 7.7); and one exhibited two blunt-force injuries, on the temporal and parietal bones; and cut marks on the occipital bones, apparently from decapitation (Rolland 1968a, 24).

Fig 7.5 Skulls from Glanum (photo Ian Armit)

Fig 7.6 Trepanned skull from Glanum (from Roth-Congès 2004, 31)

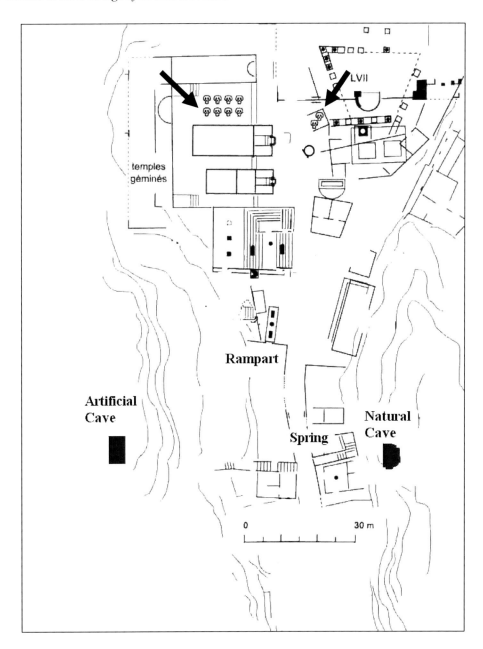

Fig 7.7 Location of skull-caches at Glanum (after Roth-Congès 2004, 25 with additions)

La Cloche c. 100 – 50 BC

Two skulls have been recovered from the main gateway of the first-century BC oppidum of La Cloche. The stratigraphy of the remains indicated that they had been displayed over the gate prior to the destruction of the site in around 50 BC (Chabot 1983, 51). The skulls are both incomplete (the base and facial bones are missing) but maxillary fragments were also discovered in the entrance, suggesting that these or other skulls were gibbeted above the gate until they fell apart. Both individuals were male, aged 30 - 40 and more than 55 - 60 years (Mahieu 1998, 64-65). One of the skulls was pierced with an iron carpentry nail (fig 7.8) like some of those from Entremont. The other skull was encased in an iron armature, which pierced the length of the skull with iron nails, and grasped the calvium (the skull minus the facial bones) in a band which ran from the occipital to the

frontal bone where it diverged into two prongs which entered the orbits (fig 7.9). A second such armature found in the vicinity, suggests that another skull was displayed over the gate of the oppidum in the same way (fig 7.10) (Chabot 1983, 50-51). From visual inspection at the Musée de Vieille Charité, Marseille, one of the skulls appears to bear two short, parallel cutmarks above the left ear. These are not mentioned in Mahieu's report, and it was not possible to obtain permission for a close examination of the remains. Nevertheless, from what could be observed, the cut marks appeared unhealed, perhaps indicating that one of the individuals had had his left ear cut off around the time of death. More cranial fragments and a mandible indicating a minimum of four individuals were found in proximity to the entrance gate. The inclusion of carpentry nails amongst the debris

108

associated with these remains has prompted the excavator to speculate on the presence of additional *'trophées'*, however, he rightly points out that these bones could just as easily have come from individuals killed during the Roman assault (Chabot 1983, 50). Discoveries of human bone, particularly in the first half of the twentieth century, very likely went unremarked unless the remains consisted of complete or near complete skeletons or crania. It may be, therefore, that the curation and display of skulls or other cranial elements was much more widespread that the above synthesis would indicate.

Fig 7.8 (top) Skull 2 (in armature) from La Cloche. Fig 7.9 (above) Skull 1 from La Cloche pierced by an iron nail. (from Mahieu 1998, 64 and 65)

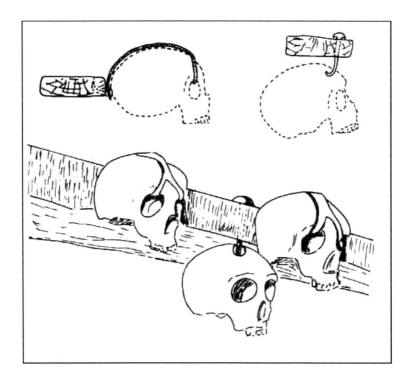

Fig 7.10 Reconstruction of the display of the skulls from La Cloche.

Two examples of this elaborate armature were found, one still containing cranial remains (Skull 2). The other skull was pierced by a nail in the manner of the skulls from Entremont. Both remains were found in wheel-ruts of the settlement entrance-way. They appear to have been suspended above the gate of the site for some time before falling into the entranceway around the time that the settlement was destroyed, in c. 50 BC

(From Chabot 1990, 120).

7.4 Human Remains

Finds of human remains which include post-cranial elements are rare in the Lower Rhône Valley; and consist, for the most part, of stray bones from settlement sites. Unlike skulls and cranial elements, these remains are too disparate to have been assigned to any particular tradition. The infant burials from Roquepertuse and Entremont, appear to have been ritualistic, however, these occur too rarely to be interpreted within the context of a cohesive custom; and, outside these kinds of burial contexts or destruction layers, stray finds of bones or skeletons are more difficult to categorise or explain.

Stray finds from Settlements

Entremont 175 – 90 BC

Partial skeletal remains comprising two skulls and the post-cranial bone of at least two individuals were discovered during the 1986-87 excavations at Entremont under the direction of Willaume, Arcelin and Congès. It was only possible to determine the age and sex of one individual: a man of at least forty years old, discovered in a house in the Ville Haute (the earlier phase of settlement: 175 – 150 BC) (Mahieu 1998, 62). It is not stated whether these remains appeared associated with either of the two known assaults on the oppidum in c. 125 and 90 BC.

Saint-Pierre c. 550 BC – AD 100 and L'Île c. 430 – 200 BC

The settlements of Saint-Pierre and L'Île in Martigues have produced fragments of cranial and post-cranial bone, which have not been published in detail. They have been found amongst otherwise ordinary domestic debris in houses and streets; and Chausserie-Laprée states that they cannot be associated with assaults, disturbed burials or the 'cult of the head' practiced around Massalia (2005, 227).

Mourre du Bœuf, Late Iron Age (date uncertain)

During a brief rescue excavation in 2004, the burial of a mature woman was discovered at the foot of the small oppidum of Mourre du Bœuf, Martigues. There were no grave goods but two iron rings closed with rivets were found at the woman's feet. These rings have been identified as slave-shackles of a type found at the Celtic site of Sanzeno in north Italy (Chausserie-Laprée 2005, 234; Duval et al. 2005).

La Cloche c. 100 – 50 BC

The skeleton of a woman, around 30 or 40 years old, was found on a bench inside one of the houses at La Cloche (fig 7.11). The woman was lying on top of a jar and Sanzeno-type slave-shackles were found nearby. The body, which had been covered by the collapsed wall of the house, was incomplete and disturbed (Mahieu 1998, 64; Chabot 2004, 107; Chausserie-Laprée 2005, 234).

The bones of one leg, which may belong to this skeleton, were found in the street near the house (Chabot 1983, 79).

Fig 7.11 Female skeleton with jar buried in a house at La Cloche. The slave chain found with these remains and the analogous burial of a woman at Mourre de Boeuf (http://pm.revues.org/index202.html) may indicate sacrifices of slaves at periods of unrest. (From Chabot 2004, 107)

A leg (tibia and femur) was discovered at the entrance gate in association with the cranial fragments described above. The leg had been pierced with a metal object, no longer extant, but which the excavator suggests might have been a Roman pilum point (Chabot 1983, 50).

Other human remains (not specified in excavation publications or in any osteological analyses) were discovered at different points around the oppidum, however, many had undergone animal disturbance. Chabot links all these remains with the assault on the settlement in around 50 BC (*ibid.*).

Infant Burials

In the above mentioned synthesis of Languedocien finds of human remains (Schwaller and Dedet 1990), burials of new-borns in house-floors figure highly. In Provence, however, such deposits are rare. One, perhaps anomalous example, from the very northern limit of the region, in Sainte-Colombe, features an unusually old child, of around six years, interred beneath a house floor of the seventh or sixth century BC. The child's head appears to have been crushed; and while it is not known if this injury was ante-mortem or post-depositional, the deposit has been interpreted both as a human sacrifice or a skull-related ritual (Courtois 1975, 24 cited in Nin 1999, 268-69). The other five Provençal deposits were all discovered in the Lower Rhône Valley, and comprise peri-natal or infant depositions comparable to those from Languedoc. The earliest date, again, to the seventh or sixth centuries BC and comprise the interment beneath house floors of a new-born infant at Saint-Blaise (Bouloumié 1984, 46-47) and an eight-month foetus at Baou de Saint-Marcel (Rayssiguier 1983, 72; Nin 1999, 269). At Roquepertuse two new-born infants were found in pits beneath house walls dating to the third century, though it has not been possible to distinguish whether these depositions were made during the period of the sanctuary (early- to mid-third century) or the oppidum (mid- to late-third century) (Boissinot 2004, 52). Finally, at Entremont, an infant was interred beneath the 'forge', an area devoted to metal-working in the second-century Ville Basse (Arcelin 2000; Chausserie-Laprée 2005, 233).

7.5 Archaeological Context

The First Iron Age

The earliest identified human remains in the Lower Rhône Valley are the peri-natal infants buried beneath house floors at Saint-Blaise and Baou de Saint Marcel. In an analysis of the Languedocien material Armit (2006) draws attention to what he sees as a divergent correlation between deposits of new-borns and deposits of adult crania. He argues that this may indicate that the two traditions were utilised to serve a similar social or ritual function. The small number and disparate contexts of similar infant burials in Provence makes the identification of such relationships less practicable. What can be noted is that these deposits occur on two of the most significant known settlements of this period. While little is known of either site in the Early Iron Age, Saint-Blaise at least, appears to have played a significant role in the southern French trade relationships with Etruria. The settlement was also around five times larger than most contemporary indigenous centres (Chausserie-Laprée 2005, 60), and surrounded by satellite oppida, strongly suggesting that it exerted some political and territorial control in the region (*ibid.*).

Third Century

Roquepertuse

The first direct evidence of practices relating to the deposition and display of skulls in the Lower Rhône Valley comes from the early third-century sanctuary at Roquepertuse. It was suggested in the previous chapter, that differences in the workmanship and morphology of niches in the Roquepertuse portico may represent an accumulation of relics from other sites in the region. Some niches were shallow, of a form appropriate for the display of facial bones such as those described by Gèrin-Ricard (1928, 4). Coignard and Coignard (1991) meanwhile have demonstrated a range of possible displayed remains based on the various other niche forms discovered at Roquepertuse, including complete, unfleshed skulls and remodelled or mummified heads with cervical vertebrae (fig 7.13).

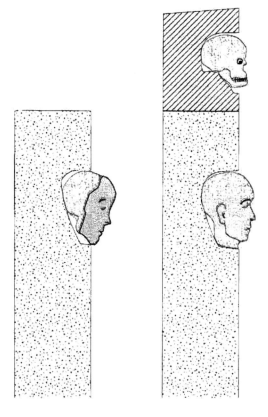

Fig 7.13 Roquepertuse niche-profiles indicating possible human remains.

Differences in depth and internal form in these niches indicates that human heads/crania in various forms were displayed in this monument.

The range of clearly functional mechanisms for holding cranial remains evidenced in such monuments, whether the 'shelves' in the Roquepertuse pillars or the internal bosses and spikes found elsewhere, strongly suggest that human heads were indeed displayed in large numbers throughout the region. Unfortunately the practice of displaying bone to the extremes of the Mediterranean climate hardly assists the survival of the remains.

(From Coignard and Coignard 1991)

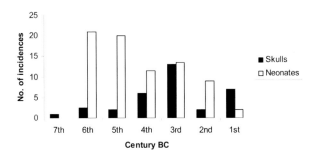

Fig 7.12 Armit (2011) sees a divergent correlation between deposits of new borns and deposits of adult crania in the Languedoc.

The figures certainly show a gradual decline of the practice of interring infants in house floors throughout the period.

In treatment, complete and interred beneath house walls and floors, these neonate burials may have a stronger relationship with the practice of animal deposits than the deposition and display of human crania in either the Languedoc or Provence

(graph from Armit 2011).

These variations may constitute a long tradition of curation leading to the accumulation of remains in differing states of deterioration (Ian Armit pers. comm.), or local variations in the ritual treatment of heads, or both. It is likely that the eventual mode of display of these 'relics' (perhaps exposing already old, curated bone to the extremes of the Mediterranean climate), resulted in a poor level of preservation. If this is the case, more fragmented elements, at Roquepertuse and, probably, at other sites, may have been either unnoticed or unrecorded in subsequent excavations. Given the prevalence of pillars with cephaliform niches, and the frequent inclusion of practical devices to assist the housing of cranial remains in such monuments, this mode of head-display must be presumed to have been much more prevalent than the current osteological evidence would indicate.

The Roquepertuse assemblage was too fragmented to allow the identification of possible trauma. Where known, the sex and age determinations of the individuals (two males aged between 35 and 45 years, a male over 50 and a possible female) may suggest that not all of these remains were the heads of warriors taken in battle in the manner described by Poseidonius. Roquepertuse does, however, produce evidence of an assault around 400 BC, shortly before the foundation of the sanctuary, and women and older men may constitute a demographic consistent with raiding. The treatment and display of the cranial remains does not appear particularly disrespectful. The heads were decorated and became, in effect, part of the ritual furniture of the sanctuary; although, as discussed in the following chapter, the treatment of enemy heads as valuable and honoured totems, as in the Amazonian tsantsa rituals (fig 7.15), is not unknown in the ethnographic record (e.g. Zikmund and Hanselka 1963). Nevertheless, niche pillars for the housing of such remains occur on settlements where they cannot always be linked to specific episodes of destruction.

It may be more likely that the 'skulls' from Roquepertuse represent a tradition of head-curation practiced more widely in the surrounding countryside, at a time when the indigenous populations of southern France were establishing more formal and permanent occupation of the territory. The early-fourth century BC saw the emergence of more standardised settlement forms, and more long-term and formalised trade relationships with the increasing number of Greek foundations (and, probably, with other indigenous communities as well).

The socio-political background of the Roquepertuse sanctuary and the themes expressed in the iconography of this period, outlined in chapter six, appear instead to indicate the development of an ancestor cult. The preservation and/or display of heads may have been utilised to reinforce the territorial claims of communities, and the political claims of emerging elites. It was proposed in the previous chapter that the destruction of the sanctuary in the mid-third century constituted a 'foundation deposit' of the collective ancestors of the

surrounding territories, and the birth of a new confederation which founded the oppidum. The use of human remains in this manner is known in the ethnographic and historical record. During the Ilgonquin Feast of the Dead, for example, the dead of several tribes were exhumed, displayed on a scaffold and subsequently interred collectively, in order to formalise alliances (Speal 2006). Whether the facial bones from Roquepertuse originated with the 400 BC assault, or were the curated relics of 'ancestors'; they appear to have been used in a ritualised and political way, rather than as the crude 'spoils of war' described by Poseidonius.

As mentioned above, the two third-century infant burials from Roquepertuse cannot be assigned definitely to either the period of the sanctuary or of the oppidum; however, the excavator believes them to be associated with the domestic rather than the public ritual sphere (Boissinot 2004, 52). In Roquepertuse (unlike other Provençal settlements yielding such deposits) similar depositions of small animals (pigs and sheep) were also placed beneath house walls (*ibid.*). In this respect, the Roquepertuse infant depositions bear the greatest similarity in Provence to the Languedocien tradition which is generally interpreted as a form of foundation deposit (Dedet and Schwaller 1990).

Second Century

Buffe Arnaud

The majority of surviving cranial remains date to contexts of display or deposition at the end of the period, in the second or first centuries BC; the earliest of which may be the two skulls from Buffe Arnaud. These remains were discovered in the levels immediately prior to the 125 BC destruction level around the gateway. Given the difficulty of removing the mandible from a newly defleshed head (Chris Knüsel pers. comm.) the absence of mandible fragments suggests that the heads had been either curated, or deliberately defleshed, prior to this display, presumably in order to comply with some convention of treatment and exposure.

The condition of the exposed bone (the post-mortem loss of teeth and the greater absence of the lower parts of the crania) indicate that the skulls had been hung over the gateway for some time prior to the destruction of the site. While no evidence of violent trauma could be discerned on these skulls, their display in the entrance of the oppidum, suggests a phenomenon more like the 'trophy' skulls, described by Poseidonius. The historical context of the display of these crania, however, remains unknown, and they cannot be associated with any particular episode of warfare. They may conceivably have pre-dated the concentration of (apparently) Roman campaigns around 125 BC by some time. No other direct evidence for aggression has, as yet, been identified at the site, and the skulls from Buffe Arnaud may originate from any period throughout its century-long occupation between 225 and 125 BC.

Entremont

Entremont Ville Haute 175 – 150 BC

The earliest human remains found at Entremont may be the incomplete skeletons discovered in the initial Ville Haute oppidum, though they may not be contemporary with this phase of settlement (175-150 BC). Despite their relative integrity, these remains appear to yield no evidence for the presence of warfare other than their presence on the site. They are not linked stratigraphically to either the 125/23 BC or the 90 BC conflict, and they are not recorded as revealing any discernible sign of trauma.

Entremont Ville Basse 150 – 90 BC

The baby buried beneath the Ville Basse 'forge' also presents some puzzling features. These remains represent the only infant deposition outside the domestic sphere in Provence, and the connection with a metalworking area may represent a more overtly sacrificial form of 'foundation offering' than, for example, the interment of a deceased infant beneath a family home. The anomalous nature of this burial makes anything more than a speculative interpretation of its meaning difficult; however, there appears to be nothing thematically or contextually 'warlike' about the deposition.

Entremont Ville Basse 125 – 90 BC

The skulls from Entremont have yielded no discernible evidence of peri-mortem trauma; and, regarding those individuals for whom an age at death could be determined, over half seem to have been between 40 and more than 50 years. This may not immediately suggest that these remains represent trophies taken from enemy warriors killed on the battlefield, but, as mentioned above, there is nothing to suppose an age-limit (or, indeed, a sex-restriction) on those killed in raiding or assaults on settlements. Niche pillars have been found at the site, perhaps indicating that 'ancestor' heads were displayed on the site, perhaps prior to the foundation of the early oppidum in around 175 BC. If this is the case, at least some of the fragmentary crania in the Entremont assemblage may have come from this earlier and more ritualistic context.

The pillar engraved with heads, and the *bloc aux épis* dating to around the sixth century would support the theory that the site was a significant focus for head-related ritual practices prior to the foundation of the oppidum in the second century. Nevertheless, the six skulls with perforations for nails or large nails still embedded in the bone, strongly suggest trophies such as those at Buffe Arnaud; and it has been proposed that these were attached to the wooden posts along the front of the post-125 BC hypostyle hall, around which the skulls were found. In this light, the stratigraphic and historic context of these remains is interesting. These skulls appear to have been displayed after the first Roman assault in around 125/23 BC, on the facade of a public building which faced a processual entrance-road, metalled with the statues of mid third-century indigenous elite (see fig 6.45). If, after the initial 125/3 BC assault, Entremont came under Roman control, there can have been no more potent symbol of the regime change, than the heads of the deposed ruling elites overlooking the iconography of indigenous, aristocratic power.

Glanum

The skull deposits from Glanum are broadly contemporary with those of Entremont, and come from a similar historical context; dating to a period between an initial Roman assault in around 125 BC and the ultimate destruction of the site by Roman forces in around 90 BC. There is little published data on the cache of eight skulls beneath the Roman temple; however, certain inferences may be made regarding how these remains could relate to warlike or violent events. Firstly their apparent youth at the time of their deaths fits more easily with the interpretation of these heads as war-trophies taken in battle. Secondly, Rolland's observation that they bore no evidence of trauma or beheading can be discounted due to the very cursory examination of these remains. The first two cervical vertebrae (indicating decapitation) may yet be inside the earth-filled skull cavities of some of these skulls (Linda Fibiger pers. comm.), and the presence of mandibles suggests that the heads were at least 'fresh' if not fully-fleshed, at the time of their deposition. Finally, the deposition of eight heads, all in a similar state of decay, and all of a similar age, may suggest that these remains represent individuals who died at around the same time, rather than an assemblage of curated heads.

The two skulls from a separate deposit appear to have undergone rather different peri-mortem treatments to those above. The evidence for trauma indicates that, at least one of these individuals met with a violent death; and, in light of this, the unhealed trepanation (if correctly identified) suggests an attempt at medical intervention which belies an interpretation of an enemy trophy-head. If these are the heads of warriors, injured in battle and taken back to the settlement to undergo medical treatment, then the evidence for decapitation, which Rolland recorded on the injured skull, may be read as part of the mortuary treatment; an interpretation which has implications for the other skulls from Glanum.

The (admittedly limited) data regarding all ten of these skulls suggests that they may very well represent evidence for warfare. Nevertheless, none of the skulls were apparently displayed as trophies. Neither deposition appears overtly disrespectful and at least two individuals seem to have been treated with care. Their contexts (beneath ritual structures from the period of 'Romanisation' between c. 125 – 90 BC) imply that they may originate from the 125 BC Roman assault on the settlement. If this is the case, they may represent Roman or indigenous soldiers, or perhaps, a mixture of both, interred separately but respectfully in the interests of political stability. As may be remembered from the previous chapter, indigenous iconography continued to be incorporated into, or displayed within, the numerous Roman monumental buildings established in Glanum after 125 BC.

The intricacies and pit-falls involved in disentangling levels of religious syncretism as outlined by Webster (1997); fall beyond the scope of this study. Nevertheless, it is well attested that Rome frequently made use of indigenous religious symbols to gain a more stable form of control than military force alone could achieve (Woolf 1995; Webster 1997). It may be that Entremont and Glanum both experienced an inversion of their iconography as regards the human head. The architectural changes apparent at Glanum may indicate a (temporary) acceptance of some aspects of Roman cultural control, borne out by the respectful interment of heads beneath Roman ritual structures, which, might be seen as approximating a kind of warrior burial.

At Entremont, however, the nailed heads suggest a slightly different scenario. The settlement remained largely indigenous in its architecture subsequent to the 125 BC assault, however, the symbols of traditional indigenous power were destroyed. Given the specific historical context of this development, it would appear that the public display of severed heads, and smashed and trampled sculpture, indicated either Roman 'muscle-flexing', or a new, native regime allied to Rome, rather than an indigenous display of either enemies or ancestors.

First Century

La Cloche

Of all the post-cranial remains in this analysis only those from La Cloche can be definitely associated with warfare, both by context (stray bones in the destruction level) and, perhaps, directly (the pierced leg in the entrance). The female body found in a house is unusual in its completeness, and may indicate the suddenness of the attack; however the slave-chain found on the floor nearby, suggests that she was unable to escape. The analogous discovery at Moure de Bœuf (http://pm.revues.org/index202.html) suggests that these deposits may have been sacrificial.

The rather advanced ages of the two skulls from La Cloche, particularly the individual aged 50-60 years at the time of death, may not suggest warrior status, however, all other indications are strongly suggestive of aggressive decapitation and display. The two iron armatures seem designed to deliberately disfigure and 'de-humanise' the heads; and the apparent mutilation of one (which may have had its ear cut off), and the finds of mandible fragments in the ditch below the gate, may indicate that they were fully-fleshed at the time of their suspension.

The context of these heads, both in the specific historical context of the settlement, and in the regional history of the Lower Rhône Valley, also appears to situate these heads in an environment of conflict. The heads were displayed after 100 BC and prior to the violent destruction of the settlement in around 50 BC. This occurred in the final decades of Rome's 'pacification' of the Provence, nearly two generations after the fall of the Saluvian centres, when La Cloche appears to represent the last significant, indigenous stronghold in the region.

In consideration of their ages at death, the La Cloche skulls may represent political or military leaders rather than trophies taken directly from the battlefield. Given the precarious political position of La Cloche at this time, caught between Massalia and Rome, these individuals could have been almost anyone, including perceived traitors or failed military leaders from within the community.

7.6 Patterns of Warfare from Osteological Remains

Because of the highly ritualised nature of many of these deposits, certain phenomena, particularly those related to the removal, curation and display of human heads, must be examined individually and contextually, as possible ritual or mortuary treatment, and not simply assumed to be an act of aggression. Osteological evidence for warfare is presented below (table 7.1). This includes indisputable evidence for violence, that is, 'trauma', (which is found only on the skull from Glanum and on the skull, the female skeleton and the leg bone from La Cloche).

Indirect osteological evidence for conflict (comprising the apparent cases of war trophies found at Buffe Arnaud, Entremont and La Cloche) is also presented, as are the skull depositions from Glanum, (which may be war trophies, but are proposed here to represent warrior burials of the in-group rather than the enemy).

The facial bones from Roquepertuse present some problems of categorisation, as the elaborate treatment of 'heads', and their apparently honorific display, may disguise an initially violent act. However, there is little to suggest that any cranial fragments from Roquepertuse were the direct product of warfare, or displayed as the iconography of a society involved in recurrent warfare. For these reasons, the interpretation of Arcelin and Rapin (1992; 2003) is accepted, and these remains are viewed as the relics of an ancestor cult. The possible trophy skull found near the entrance of Roquepertuse is not included as it is stratigraphically unreliable, and may just as easily have come from the portico.

Finally, the infant burials from various sites are included due to their possible relationship with settlements involved in head-display and centres of indigenous political significance (with the caveat that these remains more than any are, perhaps, likely to have been overlooked in older excavations).

Stray finds of human bone, such as the two individuals from Entremont and the bone from Saint-Pierre and L'Île are not included, as they are difficult to categorise as either direct evidence for violence, or as a warlike manipulation of human remains.

	St. Blaise	Baou de St. Marcel	Roque-pertuse	Buffe Arnaud	Entremont	Glanum	La Cloche
7th / 6th C	IB x 2	IB x 2					
5th C							
4th C							
3rd C			IB x 2 AC x 2				
2nd C				WT x 2	IB x 2 WT x 2		
1st C						WB x 10 T	WT x 2 T x 2

Table 7.1 Evidence for Violence and Warfare in the Human Remains of the Lower Rhône Valley.

T = trauma
WT = war trophy
WB = warrior burial
AC = ancestor cult
IB = infant burial

Osteological evidence for warriors, war trophies and for trauma appears only in the second and early first centuries BC.

The apparent preponderance of both direct and indirect osteological evidence for warfare probably reflects a level of differential survival (particularly as regards heads displayed in niche pillars) and a concentration of archaeological exploration at the larger, and therefore, later oppida.

7.7 Conclusion

As we have seen, direct and rhetorical indications of warfare are apparent only from the end of the second century to the middle of the first century BC. Comparisons with the documentary evidence (table 7.2) indicates that the osteological evidence fails to reflect even many of the later and comparatively well documented historical accounts of warfare in the region. Osteological evidence for warfare also fails to reflect the apparently intense period of warfare in around 200 BC indicated by the settlement evidence (table 7.3). These shortcomings may of course be the result of poor survival and the failure to, as yet, identify indigenous funerary or mortuary practices in southern France.

Human remains, from southern France in general and the Lower Rhône Valley in particular, are often fragmentary and many, such as the skulls from Buffe Arnaud, appear to have undergone curation and modification. Perhaps for this reason, the osteological record is most informative in comparison with the iconographic evidence (table 7.4).

The niche pillars discussed in the previous chapter suggest that the use of heads as ancestral relics, and, perhaps, social and territorial markers, began during the period of urban consolidation in the late-fifth /early-fourth century BC.

The iconography of heads in contiguous rows and inverted in *têtes-béches* motifs, however, which pre-date these ancestral porticos, can be traced to the Late Bronze Age and perhaps even earlier. It may be that this earlier iconography of death, rebirth and fertility, has its corresponding 'osteological iconography' in deposits of infants beneath domestic structures and adult cranial and post-cranial remains in caves and shafts. It seems likely that further evidence, such as that from Saint-Blaise, Baou-de-Saint-Marcel (infant deposits) and Aven Plérimond ('shaft' or cave burials) has been overlooked in the excavation strategies employed in Mediterranean France until quite recently.

Century	Historical Evidence	Osteological Evidence
>6th		**St Blaise** IB x1 **Baou de St. Marcel** IB x1
5th		
4th	390 BC siege of Massalia	
3rd	General unrest	**Roquepertuse** IB x 2 AC x 2
2nd	154 BC Q. Optimus's Campaign 123 BC defeat of Saluvian capital and foundation of Aix 106 BC Caepio's Campaign 102 BC Marius's Campaign	**Entremont from c. 150 BC** IB x1 **Buffe Arnaud after c. 225 until c. 125 BC** WT x 2 **Entremont after c. 125 until c. 90 BC** WT x 6 c. 14 other skulls **Glanum between c. 125 – c. 90 BC** WB x 10 Tx1
1st	49 BC Caesar's siege of Massalia	**La Cloche c. 50 BC** WT x 2 T x 2

Table 7.2 Comparison between the osteological and documentary evidence for violence and warfare.

Infant burials and remains associated with apparent Ancestor veneration at third century Roquepertuse, are included for comparison with more overtly warlike remains.

The discrepancy between the archaeological and documentary evidence might be attributable to poor survival of remains and poor site prospection. Nevertheless, even within this later period, only the remains from the first assault on Entremont and the final destruction of La Cloche can be related to a specific historic account of warfare in the region.

Century	Direct Aggression	Indirect Aggression	Fragmentation (house sizes)	Communality	Osteology
6th	some	some	large	some	**St Blaise** IB x1
5th	some	some	large	some	
4th	least	least	smallest	frequent	
3rd	least	least	smallest	frequent	**Roquepertuse** IB x 2 AC x 2
Early 2^n	greatest	greatest	small	absent	
Late 2nd	greatest	greatest	largest	greatest	**Buffe Arnaud** WT x2 **Entremont** WT x 6 (known) c. 14 (possible) IB x1 **Glanum** WB x 10 T x 1
1st	greatest	greatest	largest	greatest	**La Cloche** WT x 2 T x 2

Table 7.3 Comparison of the osteological and settlement evidence.

There are no osteological remains related to the period of intense warfare noted through the settlement evidence at the beginning of the second century, though these may survive as curated remains in later assemblages.

All the overtly warlike human remains were found in contexts of late second/early first century conflicts between the indigenous populations and Roman forces. Relatively few of these remains show evidence for trauma, being predominantly 'rhetorical' expressions of warrior culture: trophies and/or burials.

Century	Iconographic Evidence	Osteological Evidence
< 6th	Imagery reflecting fertility and cycle of life, death and re-birth	**St. Blaise** IB x1 **Baou de St. Marcel** IB x1
5th	Warrior statues = ritual power warriors association with libations pillars with niches (ancestor cult)	
4th		
3rd	Curated warrior statues = ritual power Pillars with niches: Heads = ancestors Warrior statues = political & military power: Heads = subjugated	**Roquepertuse** IB x 2 AC x 2
2nd	Curated 3rd century Warrior statues = political and military power Heads = subjugated Curated 5th century warrior statues = ritual power	**Buffe Arnaud** WT x 2 **Entremont** WT x 6 c. 14 skulls IB x1 **Glanum** WB x 10 T x1
1st	Period of Gallo-Roman iconography: Tarasque of Nove, Sphinx of Orange etc.	**La Cloche** WT x 2 T x 2

Table 7.4 Comparison of osteological and iconographic evidence.

The table reflects a close relationship between iconography and human remains throughout the Iron Age. The early centuries have produced little human bone, however, the placement of infant foundation deposits seems an appropriate counterpart of the themes of fertility, life, death and rebirth, expressed in the iconography.

The seemingly more honorific treatment of the human head is first seen in the early third-century portico at Roquepertuse, but may reflect practices carried out from the end of the fifth century. This suggests that the adoption of the head as an ancestor relic coincides with the creation of figures combining the attributes of warriors and 'priests'.

After the destruction of the sanctuary at Roquepertuse, both heads and warriors take on more literal roles. The indigenous sculpture portrays warriors as politically and militarily victorious, while heads become either the subjugated enemy or the honoured soldier.

The iconographic and osteological remains related to the display and depiction of severed heads in southern France were discovered long after the literary allusions to 'Celtic headhunting' were already common currency. These cranial remains have usually been viewed as the heads of enemies killed in battle (Salviat 1993) while the more elaborate nature of some of the manifestations of head-related practices has led to the interpretation of coeval traditions which denigrated the heads of enemies, and revered those of ancestors (Arcelin and Rapin 2003). Despite indications, in literature (e.g. Silius Italicus XV.807-23 (Punica)) and iconography (e.g. Trajan's column) that the Romans themselves very often took and displayed the heads of their enemies (fig 7.14), evidence of the 'head cult' in southern France has only ever been attributed to the Gauls.

Fig 7.15 Scene from Trajan's Column showing (partially destroyed) depiction of a Roman soldier holding the decapitated head of his Dacian enemy in his teeth.

(Scene 24, CD 4, image 89 of the McMasters Trajan's Column Database. Photograph: Peter Rockwell)

The removal, curation and display of human heads is a recurring phenomenon in the archaeological, historical and ethnographic record; and the range of possible meanings attached to these customs may be as diverse as the societies which practiced them. In the Neolithic Levant, for example, the decoration of deposited or displayed skulls, and a body of iconography depicting decapitation (Kjuit 2001, 80-99) have resulted in attempts to identify how this custom reflected, or impacted on, the social and political lives of the societies in which it was practised (Kjuit 2000; Goring-Morris 2000). Likewise in southern France, finds of displayed or curated skulls and concentrations of iconography related to *têtes coupées*, have highlighted the need to interpret these phenomena as a coherent practice rather than as a random by-product of conflict.

Unlike in the Levant, however, the osteological and lapidary remains first discovered in southern France in the nineteenth and early-twentieth century, appeared with a ready-made cosmology:

> *there is also among them the barbaric and highly unusual custom...of hanging the heads of their enemies from the necks of their horses when departing from battle, and nailing the spectacle to the doorways of their homes upon returning...The heads of those enemies that were held in high esteem they would embalm in cedar oil   Strabo (4.4.1).*

While not necessarily shedding much light on the patterns of warfare in the region, a contextual study of the osteological evidence in the Lower Rhône Valley suggests that the interpretation of displayed heads as a correlate of war or the war-like practices of a particular group, may be misleading.

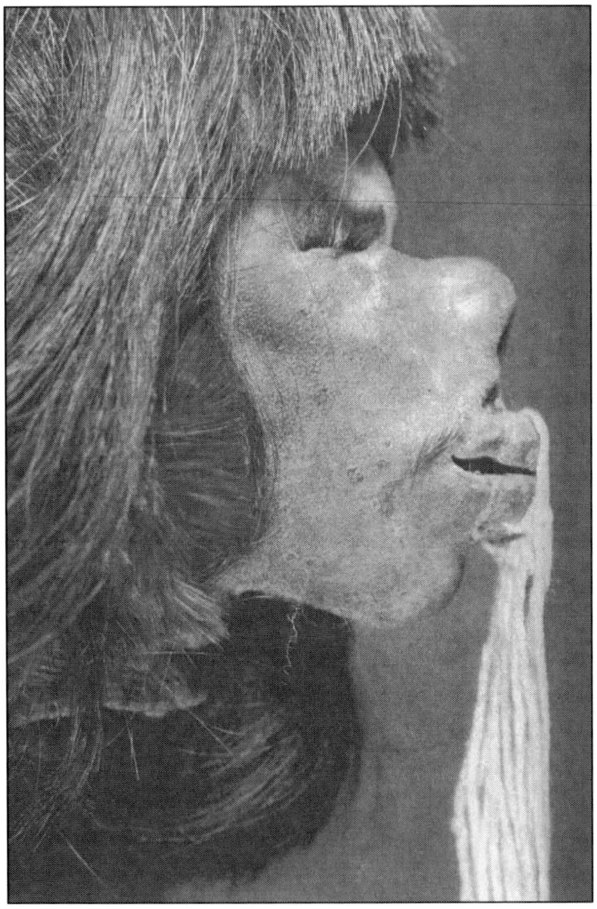

Fig 7.16 Southeast Asian Iban Head.

Heads taken during the nineteenth century still displayed in an Iban longhouse.

These heads illustrate the difficulty of isolating the socio-political context of such practices.

The proliferation of head hunting and display in the nineteenth century was a direct result of a 'civilised-tribal zone' contact with state-level European markets.

The heads indicate transformations in social complexity and the Iban world view caused by this contact but the expression of this change (intensification of warfare and headhunting) is wholly indigenous.

Photograph by Janet Bell

Fig 7.15 Amazonian tsantsa head.

These heads are taken as part of aggressive, predatory headhunting practices but are ritually transformed into elaborately decorated, powerful and valued talismans.

The ultimate use and display of human heads or cranial remains may be a poor indicator as to whether or not they originated from acts of violence or reverence (From Zikmund and Hanselka 1963, plate 16 between pp 240 – 241)

8.1 Introduction

One of the primary aims of this research has been to identify patterns of conflict throughout the Iron Age in southern France with a particular focus on the area around the Lower Rhône Valley. This region has long been associated with a corpus of apparently warlike iconography and what appeared to be the osteological vestiges of the war-related headhunting practices recorded by ancient Roman and Greek sources. The previous four chapters have analysed these and other possible archaeological indications of warfare. The usefulness of these strands of evidence (the documentary sources, the ordering and defence of settlements, iconographic themes and osteological remains) as a means of recognising conflict is assessed in the following chapter. For now it is sufficient to note that, while the different strands of evidence sometimes conflict, and often highlight different aspects and episodes of warfare; when considered in tandem the results of these analyses display a good deal of consensus. This chapter begins, therefore, by outlining the patterns of warfare and violence throughout the Iron Age in the Lower Rhône Valley.

The first two chapters summarised the very different approaches to warfare in archaeology and in social anthropology. Archaeologists have often viewed warfare (particularly among societies with lower levels of social organisation) as a by-product of social interaction rather than a social process in its own right. Groups may have plundered their neighbours for revenge or to gain cattle or slaves, while individual warriors could achieve prestige and personal wealth through raiding (e.g. Ross 1986, 40-53). Such violence was seen as episodic and without consequences, and any material gains were 'incidental'. Warfare was rarely seen as a means of consolidating territory or resources at a polity-level (Ferguson 1984, 23).

Among more complex societies, the motivations and consequences of going to war have been afforded greater deliberation (Keeley 1996, 11). For example, the extent to which Rome was tied into systems of slave-trading or imperial conquest by its early wars in Italy and Mediterranean Europe has, and continues to be, discussed (e.g. Cunliffe 1978, 63; Taylor 2001). In general, however, conflict or violence in the archaeological record has tended to be documented as a series of isolated and episodic events. By contrast for over a century social anthropologists have considered warfare, not as an isolated series of events, but as a normal social process. Warfare is seen as having far-reaching consequences for social structure, cultural expression and political organisation; and can be both a catalyst and an ongoing mechanism in these arenas (e.g. Harrison 1993; Deflem 1999).

The second part of this chapter, therefore, examines the changing levels of warfare in the Lower Rhône Valley in relation to the anthropological models discussed in chapter two, specifically those which deal with the causes and consequences of warfare. This section draws on some of the most relevant of these models in order to identify the extent to which conflict in the region was driven by external colonial forces or internally-driven indigenous mechanisms; and the effect of this conflict on indigenous socio-political organisation. In doing so, this study offers, not only a new interpretation of warfare in the Lower Rhône Valley during the Iron Age; but a case study for how warfare might be integrated into archaeological theories concerning the mechanisms of social change.

8.2 Patterns of Warfare

Considered in tandem (table 8.1), the various strands of evidence for warfare appear to reveal steady levels of warfare throughout the early centuries of urbanisation. These levels rise sharply around the fourth and third centuries BC, then peak in the second century, before falling off again in the first century. The more conventional forms of evidence (historical sources, direct evidence for aggression and indirect evidence of defence) appear to be appropriate for identifying large-scale, externally directed warfare. Patterns of warfare compiled using only these forms of evidence suggest that the greatest period of warfare took place in the late-second century BC (the time of the Roman intervention).

More exploratory forms of evidence for warfare (e.g. socialisation for mistrust, demonstrated in the fragmentation of communities into small households and an avoidance of communality) appear to suggest higher levels of conflict around the third and early-second centuries BC. This may suggest that indications of fear and mistrust have the potential to detect small-scale, internal warfare not identified by conventional forms of evidence. By the same token, the lack of evidence for mistrust in the proto-state settlements of the later second century may indicate that socialisation for mistrust is not a reliable indicator of warfare in societies with more complex social structures. This interpretation would reflect the societies included in the original cross-cultural study (Ember and Ember 1992 and pers. comm.) which were mostly non state-level. In the case of the southern French Iron Age, these results could suggest that the concentration of direct evidence for aggression around 200 BC was the culmination of around a century of small-scale, internal conflict in the region around Marseilles.

Considered together, all the strands of evidence suggest that levels of warfare in the Lower Rhône Valley remained broadly constant during the sixth and fifth centuries, began to rise during the fourth and third centuries, and reached a peak in the second century BC.

Table 8.1 Patterns of Conflict between the 6th and Early 1st Century

1 = no evidence for warfare

- no historical references to conflict

- no conventional evidence for warfare (episodes of direct or indirect aggression)

- no warlike iconography (no statues or engravings of warriors)

- osteology - no osteological evidence for warfare

- fragmentation - larger than average household sizes

- communality – all kinds of communal space (unbuilt space, shared utilitarian space, formally defined public space)

2 = little evidence for warfare

- vague references to aggressive behaviour

- one – two examples direct aggression (conventional evidence for warfare)

- iconography of 'the warrior' with ritualistic overtones

- osteology - evidence for head curation / veneration

- fragmentation - average household sizes

- communality – less formal kinds of communal space (unbuilt space, shared utilitarian space)

3 = frequent evidence for warfare

- references to specific battles

- three-four examples of direct aggression (conventional evidence for warfare)

- iconography of 'the warrior' with militaristic overtones

- osteology - evidence for head curation with aggressive overtones (trophy heads)

- fragmentation – smaller than average household sizes

- communality – no utilitarian shared space but evidence for unbuilt space

4 = recurrent warfare

- references to specific campaigns

- five or more examples of direct aggression (conventional evidence for warfare)

- iconography of 'the warrior' in postures representing combat

- osteology – unambiguous evidence for aggression (trophy heads) and violence (trauma)

- fragmentation – smallest household sizes

- communality – no communal or open spaces

Table 8.1a	
6th Century	
Historical evidence for warfare	2
Conventional evidence for warfare	2
Fragmentation	1
Communality	3
Iconography	4
Osteology	1
Total	**11**
Total minus documentary and conventional evidence	**9**

Table 8.1b	
5th Century	
Historical evidence for warfare	1
Conventional evidence for warfare	2
Fragmentation	2
Communality	3
Iconography	2
Osteology	1
Total	**9**
Total minus documentary and conventional evidence	**8**

Table 8.1c	
4th Century	
Historical evidence for warfare	3
Conventional evidence for warfare	2
Fragmentation	3
Communality	3
Iconography	2
Osteology	1
Total	**12**
Total minus documentary and conventional evidence	**9**

Table 8.1d	
3rd Century	
Historical evidence for warfare	1
Conventional evidence for warfare	2
Fragmentation	4
Communality	2
Iconography	3
Osteology	3
Total	**13**
Total minus documentary and conventional evidence	**12**

Table 8.1e	
Early 2nd Century	
Historical evidence for warfare	1
Conventional evidence for warfare	4
Fragmentation	3
Communality	4
Iconography	4
Osteology	4
Total	**16**
Total minus documentary and conventional evidence	**15**

Table 8.1f	
Late 2nd / Early 1st Century	
Historical evidence for warfare	4
Conventional evidence for warfare	4
Fragmentation	1
Communality	1
Iconography	4
Osteology	4
Total	**14**
Total minus documentary and conventional evidence	**10**

Table 8.1g						
	6th Century	5th Century	4th Century	3rd Century	Early 2nd Century	Late 2nd / Early 1st Century
historical & conventional forms of archaeological evidence only	4	3	5	3	5	**8**
all forms of evidence	11	9	12	13	**16**	14
experimental forms of archaeological evidence only	9	8	9	12	**15**	10

Table 8.1g The table shows different patterns of warfare using conventional forms of evidence (documentary evidence, and evidence for direct assault and defensibility), a more contextual form of evidence (utilising a combination of the strands of evidence employed in this study) and more experimental forms of evidence alone (i.e. Evidence other than than documentary evidence, evidence for assault and defensibility).

The highest levels of warfare are indicated in bold. Acute rises in levels of warfare (rises of three or more 'points' are shaded.

The traditional forms of evidence (historical and conventional archaeological evidence) appear to be appropriate for identifying large-scale, externally directed warfare. Patterns of warfare compiled using only these forms of evidence suggest that the greatest period of warfare took place in the late second-century BC (the time of the Roman intervention).

Other, more experimental strands of evidence for warfare and patterns utilising all forms of evidence, suggest higher levels of conflict around the third and early second centuries BC. This may indicate that evidence for fear and mistrust has the potential to identify small-scale, internal warfare. In the case of the southern French Iron Age, it could suggest that the concentration of direct evidence for aggression around 200 BC was the culmination of a prior, long-running period of conflict in the region around Marseilles.

8.3 Interpretations of Warfare

Recent investigation into the proto-historic period of southern France has moved away from earlier preoccupations with levels of Hellenisation or Romanisation. Instead, research has come to focus on how increased social complexity may have been based on trade networks and production centres which were instigated and controlled by the indigenous communities themselves. This is not to say that the influence of the Greek colonial presence has been disregarded. In the arenas of trade, production and social organisation, the Phocaeans are still considered to play a significant role in

enticing, shepherding or forcing indigenous strategies. For example, Garcia (very reasonably) suggests that the expansion of certain colonial centres, such as Arles, Emporion, Agde and Massalia itself, in the fourth century, may have influenced the move of indigenous agglomerations to the agricultural hinterlands of these Greek sites, in order to benefit from their commercial networks (2004, 88-89). Unlike patterns of trade or socio-political development, however, warfare in the region has been dealt with in a much less nuanced and integrated manner. As discussed at the end of chapter five, current thinking regarding warfare in southern France has been based on finds of Greek weaponry in destruction layers at indigenous sites, and early textual references to conflict between the Phocaeans and the native populations, leading to a near-exclusive focus on inter-ethnic conflict.

The review of recent anthropological studies of warfare in chapter two, suggests that relationships between colonialism, society and warfare may be more subtle and complex than the rather simplistic 'inter-ethnic' interpretation currently offered to explain warfare in southern France would imply. As already discussed, the broadly delineated approaches to warfare in anthropological studies (materialist, cultural and biological) allow for a good deal of convergence, and are largely a matter of emphasis on one or other factor. Of these approaches, the materialist interpretations, in particular the interpretations formulated within the materialist sub-set of 'cultural-ecology' theory, may present the most widely-applicable models for explaining the causes and consequences of warfare. These theories concentrate on the relationship between the natural environment and the way in which subsistence and social organisation is ordered (Ferguson 1984, 36-37). In this way, internal factors (that is, indigenous social processes) and external factors (the 'civilised-tribal zone' created by contact between groups of unequal social complexity) are both taken into account within the specific cultural and topographical landscape. According to cultural-ecological approaches, both internally and externally driven conflict is motivated by materialist considerations of competition over resources. The patterns of warfare carried out in response to this competition, however, would depend on cultural factors, notably social complexity and political organisation. The following discussion brings some of these perspectives especially that of Earle (1997) discussed in chapter two, to bear on the southern French material.

8.4 Warfare and Political Evolution in the Southern French Iron Age

Sixth and fifth Centuries BC: Warfare and the Emergence of 'Proto-Urbanisation'

Urbanisation in southern France began around the end of the seventh century BC (Dietler 1997, 313-14). The social processes which underlay this burgeoning of agglomerated settlement had already been set in train prior to the Greek colonisation in 600 BC (Gras 2004). Greek and Etruscan traders had been active in the Mediterranean for at least a century (*ibid.*). Saint-Blaise, tellingly, pre-dates Massalia by perhaps twenty-five years (Arcelin et al. 1983, 138-39; Bouloumié 1990, 33-36), and it is largely through finds of seventh-century Etruscan pottery that many of the smaller indigenous occupation sites of this period have been recognised (Dietler 1997, 310; Chausserie-Laprée 2005, 44-45). The earliest known indigenous enclosed settlements occur in Martigues, the area south and west of the Étang de Berre. Urbanisation in this area appears with a seemingly fully-formed settlement hierarchy, of which Saint-Blaise is generally believed to be the earliest principal centre (Arcelin et al. 1983; Bouloumié 1990; Gras 2004).

Very little is known about the Gaulish settlement at Saint-Blaise (founded around 625 BC), as subsequent Greek and Roman occupation has obliterated much of the early levels, and until the 1980s, excavations were concerned solely with these remains. However, the mass of its earliest ramparts and the huge area they enclose (5.5 hectares) (Chausserie-Laprée 2005, 60), suggest that Saint-Blaise was the principal centre of a number of indigenous foundations, such as the oppida of Castillon and Castellan, established around the fertile banks of the 'zone of lagoons' (*ibid.*). To the south, the coastal *éperon barré* of Tamaris (established around 600 BC) was also impressive, though smaller and of short duration. Tamaris could have housed at least a thousand inhabitants and the size and complexity of its ramparts are unparalleled in the southern French Iron Age (Duval 1999, 106; Chausserie-Laprée 2005, 67-69; Duval 2005). In addition to these major sites, numerous other small settlements, such as L'Arquet (Chausserie-Laprée 2005, 28) and Fourques-a-Chateauneuf (Chausserie-Laprée 2005, 42) or fortified farmsteads such as Coudounèu (Garcia 2005, 91) were established in the region in the sixth and fifth centuries.

This early period of 'proto-urban' settlement was far from static, and as major centres fell, smaller agglomerations grew in power, most notably, perhaps, Saint-Pierre and L'Île, both of which more than doubled in size over the next three or four centuries. From the mid-fifth century Saint-Blaise appears to decline, and it is generally agreed that, from the fourth century, it came under Massaliote control (Tréziny, H. 1986). Chausserie-Laprée rightly states that the expansion of L'Île around this time must be seen in relation to the loss of this indigenous power base. He suggests that the rather striking absence of significant native sites in the Begudienne hills (the large area which lies between the Caronte Channel and the Zone of Étangs) may indicate a movement of indigenous populations from the vicinity of Saint-Blaise to L'Île (2005, 45) (see Fig 5.1). It is interesting that during this transition in the native power base, Saint-Blaise reveals no evidence for assault, and that L'Île was attacked in 430 BC, before it was even fully established. This may simply reflect the dearth of knowledge about the early phases of Saint-Blaise generally, but it also suggests two different strategies at work in the acquisition of power.

The other major succession of power in this period can be seen in the fall of Tamaris and the subsequent foundation and rapid expansion of Saint-Pierre both of which take place around 550 BC. There is a marked incongruity regarding the oppidum of Tamaris. The fortifications are spectacularly sophisticated for any period in the southern French Iron Age. The settlement layout, however, with several contemporary zones of different architectural planning, suggests an absence of central organisation. Saint-Pierre, conversely, starts out around three times smaller, but with a palpable sense of centralised control, both in its settlement layout, which is very regular for such an early site; and in its pivotal place in the wider settlement hierarchy, with its series of smaller, strategically placed satellite settlements, such as Mourre du Boeuf and Vallon de Couest. Given this evidence of a strong political centrality, coupled with its highly advantageous location (controlling the Saint-Julien plain, the largest fertile tract of land in the Lower Rhône Valley) it is not surprising that Saint-Pierre rapidly became a major indigenous centre.

It is not apparent whether the shift in power from Tamaris to Saint-Pierre in the region to the south of the Étang de Berre was the result of direct competition, or simply the filling of an indigenous power vacuum. Both the fall of Tamaris and the rise of Saint-Pierre may have been accompanied by evidence for military threat. From their initial construction, the fortifications at Tamaris underwent repeated reinforcement and modification. The occupants of the settlement appear to have been involved in preparation against an assault, which came within two generations of its foundation.

Saint-Pierre, while apparently remaining unassailable itself, seems to have been inundated with a flood of incomers from the surrounding countryside. This influx of people (refugees?) appears to have occurred much quicker than the founders of Saint-Pierre can have planned. The first rampart was demolished not long after its construction, to allow the oppidum to be extended. However, this was soon followed by the partitioning of houses and the further spread of the settlement outwith the new ramparts. In addition, as at Tamaris, great attention appears to have been given to the defence of the site itself, with its ever-increasing ramparts; and to the protection of its wider territory, demonstrated in strategically placed outposts. Chausserie-Laprée contrasts the 'stabilité' and 'dynamisme' of Saint-Pierre with the other early indigenous settlements (2004, 45), but these qualities were not accompanied by complacency.

These initial two centuries of proto-urbanisation in the Midi recall Carneiro's (1970; 1990) and Earle's (1997)

analyses of emerging political complexity in chiefdoms, in which power bases rise and fall in cycles. In some instances, military force is not fully aligned with economic or ideological power, so that competition between polities results in success which is unstable and, therefore, short-lived. Regional power in this scenario moves from one power-base to another (e.g. Earle 1997, 191-211). This interpretation may be supported by the iconography of the seventh and sixth centuries, such as the Warrior of Lattes (fig 6.20) or the engravings on stelae at Glanum and Mouriès (fig 6.3). These depictions suggest that power was embodied in the persona of the warrior as a literal combatant: physically powerful, but with no ideological or economic power beyond his own heroism and wealth. The unequalled success of Saint-Pierre might be seen in its apparent ability to combine all three strands of power (military, ideological and economic). The oppidum appears to have dominated the only large fertile tract of land in the region of Martigues, and yielded evidence for large-scale agricultural production (Chausserie-Laprée 2005, 202). It may be remembered that Saint-Pierre also produced evidence for communal ritual practice as early as the fifth century, as well as an engraved head similar to the sixth-century pillar engravings found at Entremont and elsewhere. This early integration of different kinds of chiefly power, may have allowed Saint-Pierre to ride the ensuing vicissitudes of indigenous social re-organisation. In any case, it is one of the few early settlement sites which continued to develop without apparent interference into the Second Iron Age. The fifth-century tradition of warrior statuary, which apparently combines martial and ritual significance (Earle's integration of military and ideological power?) is rarely associated with the early power centres, west of the Étang de Berre. This iconography is found further east, in the region associated with small but stable indigenous centres which grew up in the following centuries in the Massaliote hinterland (fig 8.1).

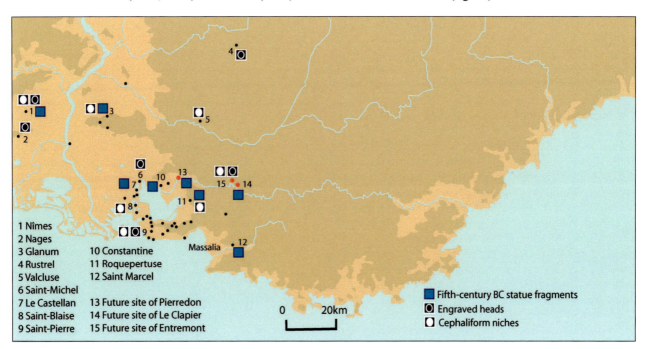

1 Nîmes
2 Nages
3 Glanum
4 Rustrel
5 Valcluse
6 Saint-Michel
7 Le Castellan
8 Saint-Blaise
9 Saint-Pierre
10 Constantine
11 Roquepertuse
12 Saint Marcel
13 Future site of Pierredon
14 Future site of Le Clapier
15 Future site of Entremont

0   20km

■ Fifth-century BC statue fragments
◙ Engraved heads
❏ Cephaliform niches

Massalia

Fig 8.1 Sixth and Fifth Century Iconography with Coeval Settlement

Fourth and Third Centuries BC: The Consolidation of Hierarchy and Centralisation

The fourth and third centuries were a period of transition in the development of Gaulish urbanisation. In these centuries the oppida of Provence became fewer and smaller. Many sites established in the sixth and fifth centuries either experienced a decline, such as Castellan, or were abandoned like Vaison la Romaine. Only Saint-Pierre and L'Île continued to develop throughout this phase, and they both adapted to the new trend of very densely occupied settlement areas (Chausserie-Laprée 2005, 46; Dietler 1997, 317). The fourth and third centuries also signalled the development of new centres in the area around Marseilles. This is the period of such oppida as Notre-Dame-du-Pitié and Teste-Nègre (Gantès 1990), most of which were quite small (mostly less than 1 hectare) and favoured upland sites overlooking agricultural land. Despite a superficial impression of dispersal and fragmentation, this period actually reveals a consolidation of urban development. Internal layouts at this time reflect a much more structured organisation, being noticeably more regular and employing a greater rationalisation of space (Garcia 2004, 80-89). Unlike the earlier indigenous settlements, these centres sometimes yield evidence for storehouses and artisanal areas, perhaps suggesting an integration of economic control with political power (cf Earle 1997).

The fourth century appears to present a lacuna as regards iconographic (or ideological) expression. It is unlikely, however, that the fifth century warrior statuary, such as that found at Glanum and Roquepertuse, could have survived by accident. It is more plausible that these figures were curated during this 'intermediate' period in southern Gaulish socio-political development. Nevertheless, the fourth century produces neither the evidence that earlier iconography was displayed, or that new traditions of iconography were developed. In other words, throughout the fourth century, while political and economic power adjusted and consolidated, the ideological expression of indigenous authority remained rooted to the fifth century themes of the chieftain as a ritual leader.

The new tradition of warrior statues created at the beginning of the third century BC appears to suggest a change in the characterisation of indigenous leadership. These figures seem more militaristic (wearing swords) and political (shown in attitudes of dominance with severed heads). The third century statues also appear to be very individualistic (Armit 2006), perhaps suggesting an attempt to legitimise new power-lineages (Arcelin and Rapin 2003). It may be that the centralised control of economic production (suggested by shared, rather than domestic, storage and production areas within settlements) and ideological power began to be integrated from this period. Military power at this time is not demonstrated through conventional forms of archaeological evidence, however, evidence for fear and mistrust (a potential indicator of small-scale, internal warfare) suggests an acute rise in levels of warfare from

the third century (table 8.1g). The Saluvian tribes of the Lower Rhône Valley, therefore, appear to bring together in the third century BC all three strands of power described by Earle (1997) as necessary for achieving, consolidating and maintaining complex chiefdoms or proto-states (fig 8.2 and table 8.2).

The iconography certainly seems to imply a period in which smaller polities were absorbed by new political centres, either voluntarily or by military force. The sculpted heads beneath the hands of the stone warriors at Entremont may have indicated the disparate territories now under Saluvian control (Armit 2006, 9). During the same period at Roquepertuse, the cranial remains once displayed singly or in small numbers outside the entrances of settlements, appear to have been collected in unprecedented numbers. The skulls, remodelled facial bones and/or preserved heads displayed in the portico at Roquepertuse, may well have symbolised an alliance, consolidating the territories in the vicinity of the Arc Valley. Chausserie-Laprée (2005, 53) notes that the kinds of agriculture practiced around the Lower Rhône Valley (possibly grain and vines from the early period, with the addition of olives, from the fourth century) provided both a potential surplus of staple foods and trade goods for the producers. Such an economy would potentially have provided a very stable power base and resembles Earle's (1997) Hawaiian case study, rather than, for example, that of the Thy, who built economic power on the trade of inter-regional prestige goods.

Little or no warfare is demonstrated through direct evidence for aggression during the third century, although Silius describes a situation of unrest between Massalia and the native populations at this time (Punica xv, 169-72). It seems unlikely that the mobilisation of troops from Rome and Massalia, and Hannibal's march through Languedoc at the end of the third century had no repercussions for the populations of the Lower Rhône Valley. They were probably not directly involved in the military action of the Second Punic War, however, Cunliffe has suggested that the Roman military presence may well have instigated internecine raiding for resources and, more particularly, slaves (Cunliffe pers comm). This is a persuasive hypothesis which, given the inroads made on the elusive subject of slavery in the archaeological record recently (e.g. Taylor 2001) certainly deserves further investigation. For the present it is sufficient to state that it does appear to reveal a period in which the small but politically and economically centralised polities in the Massaliote hinterland entered into wars and alliances of expansion. From the beginning of the third century, the iconographic and osteological evidence from Entremont and Roquepertuse suggest that smaller groups were consolidating.

In this light, the concentration of direct evidence for aggression at the end of the period, and the increased centralisation which followed, may be seen as the culmination of a long period of internal conflict and socio-political transformation among the indigenous populations.

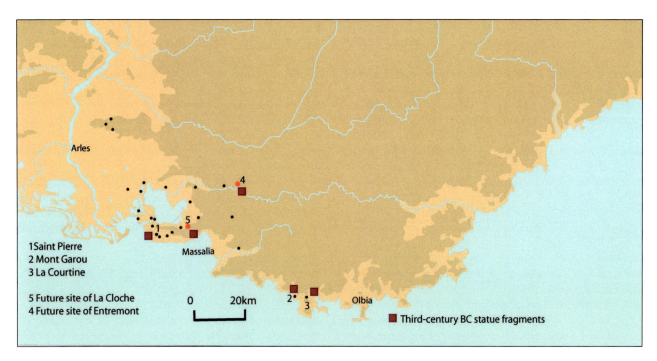

Fig 8.2 Third and Second Century Iconography with Coeval Settlement

Table 8.2a							
	6th C	5th C	4th C	3rd C	Early 2nd C	Late 2nd C	Early 1st C
Tamaris							
Saint-Pierre	20m²	20m²	8m²	8m²			
L'Arquet		16m²					
L'Île		13m²	17.3m²	10.3m²	13m²		
Notre-Dame			11m2	11m2	11m²		
Roquepertuse		24m²		21.5m²			
Teste-Nègre				8m²			
Baou-Roux					25m²	25m²	
Entremont		.			10m²	49m²	49m²
La Cloche						13.5m²	40m²

Table 8.2b							
	6th C	5t C	4'C	3rd C	Early 2nd C	Late 2nd C	Early 1st C
Saint-Blaise							
Tamaris							
Saint-Pierre	sanctuary						
L'Arquet							
L'Île							
Notre-Dame							
Roquepertuse				sanctuary	shared space		
Glanum				sanctuary			
Teste-Nègre				shared space			
Baou-Roux				shared space			
Entremont				sanctuary		shared space	
La Cloche				sanctuary			shared space

The greatest fragmentation (households under 15m²) (Table 8.2a) and the earliest and most concentrated evidence of centralised 'economic' and 'ideological' authority (storehouses, artisanal space and sanctuaries) (Table 8.2b) occur in the third and early second centuries BC. (Saint-Pierre, as well as yielding evidence of an early sanctuary, appears to have constituted the most stable known power centre of the early period).

Second Century BC: Warfare and the Rise of the Gaulish 'Proto-states'

In the second century BC native settlement underwent a dramatic change in scale, social hierarchy and political organisation. This new era is generally known as the period of the Saluvian confederation: the powerful and belligerent Gaulish alliance described by Livy and Strabo (Ebel 1976, 64-66). Settlement hierarchies re-emerged and the populations of the Lower Rhône began to congregate in much larger, fortified hilltop settlements, usually exemplified by Entremont which is generally considered to be the capital of the Saluvii (Ebel 1976, 65; Arcelin et al. 1990). These new centres were in many ways a continuation of the indigenous architectural traditions which had gradually developed over the previous four centuries. Settlements comprised domestic spaces laid out in islets of rows or otherwise abutting cells seen in the Early Iron Age agglomerations. They were laid out with the impression of centralised and overarching conception developed during fourth and third centuries, but on a much greater scale. Major settlements also included artisanal zones where metal working, food and pottery production and agricultural food processing where carried out. A striking illustration of this amplification of scale and technology can be drawn from the oil press at Entremont (Brun 1993). Most third-century settlements have produced small stone wine or (more probably) oil presses, but the Ville Basse at Entremont (*c.* 150 BC) boasted a large, purpose-built room housing a mechanised oil press adjacent to a storeroom stacked with large storage jars. In addition, these oppida made much use of monumental stone architecture in huge fortifications and public buildings, formally laid out communal and ritual spaces and other indications of increased political centralisation and social complexity (Arcelin et. al 1992, 181-82; Dietler 1997, 314-17).

While the fourth and third centuries are often seen as the period of urbanisation-proper, the new power bases which emerged at the beginning of the second century are described by Garcia (2004, 89) as true *cités gauloises*. These dramatic social changes were borne on the wake of the most sudden and widespread outbreak of violence yet evidenced throughout the Midi and, within a century, the southern Gauls were politically and militarily equipped to engage both Massalia and Rome in a long-term struggle for supremacy in the region. The aftermath of this seemingly unheralded eruption of violence, saw the substantial growth of some sites in power and complexity (e.g. Baou-Roux), the abandonment of others (e.g. Teste-Nègre) and the foundation of new centres of equally unprecedented power (e.g. Entremont). Current interpretations of this outbreak of warfare suggest the conflict was a renewal of old hostilities between the Gauls and Greeks (Chausserie-Laprée 2005, 47) or that it represented punitive attacks by Rome on behalf of Massalia (Ian Armit pers. comm. and Loup Bernard pers. comm.). Notwithstanding caveats regarding the identification of assailants through weaponry (outlined in chapter five) all assaults on settlements in the region

which are attributable to the Roman intervention, have left scatters of distinctive Roman projectiles (e.g. Entremont *c.* 90 BC, Glanum *c.* 50 BC and La Cloche *c.* 50 BC). While the participation of Massalia in the conflict around 200 BC cannot be definitely ruled out, reference to the periods of 'regression' and 'progression' described by Carneiro (1970) and Earle (1997) suggests an alternative interpretation. The obvious conclusion would appear to be that, following the period of fragmentation, or regression, in the fourth century, polities around the Lower Rhône Valley began to consolidate through alliances or under coercion by more powerful chieftains. The conflict and factionism of the third century came to an end with a period of internecine conquest warfare, resulting in the absorption of defeated territory, resources and populations into new 'proto-states' capable of taking on the nascent Roman Empire.

The clash between the Roman army and the Saluvian confederation which took place around the last quarter of the second century BC may be discussed in respect to the concept of the civilised-tribal zone, although, the opposing groups may not have demonstrated greatly divergent levels of political complexity. From the indigenous perspective, the concentration of warfare around 125 BC was a large-scale and externally directed conflict which cannot necessarily be related to the internal, political processes such as those described by Earle (1997). From the Roman perspective, the intervention was a war of state expansion and territorial conquest, after which traditional ways of indigenous life began to decline. Ferguson and Whitehead (1992, 252-54) discuss various options presented to indigenous peoples faced with a state expansion, which can be reduced to assimilation into the state, fleeing the region or the renewal of hostilities. To some extent there appears to be evidence for all three strategies in southern France. Many indigenous settlements such as Saint-Blaise or Saint-Pierre became the sites of thriving Gallo-Roman towns (Chausserie-Laprée 2005, 104-105). Walsh and Mocci (2003) interpret the sudden rise of small settlements on the slopes of Mont Saint-Victoire, at the edge of the region at the beginning of the first century BC, to an influx of refugees from the Roman wars. Finally, despite Strabo's assertion that the Romans had pacified the Transalpine Gauls, and cured them of their barbarous customs, settlements, indigenous in form and custom, (e.g. La Cloche) continued until the mid-first century.

8.5 Conclusion

Internal Warfare

Far from being isolated and episodic outbreaks of inter-ethnic hostility between Gauls and Greeks, warfare in the southern French Iron Age appears to have strongly contributed to rising levels of socio-political complexity within the indigenous milieu. Much of the conflict evident around the Lower Rhône Valley appears to correspond with patterns of warfare in anthropological studies, which describe war in tribal or chiefdom societies as a tool in the formation of more centralised political structures. Sometimes, as in the case studies presented by Carneiro (1990) greater socio-political complexity is a by-product of conquest warfare. Earle, particularly in the case of the Hawaiian Kaua'i (1979, 200-11) suggests that warfare can also be a deliberate strategy in increasing and retaining political power. The nature of socio-political change in southern France suggests that much of the conflict, particularly in the period between 600 BC and the Roman intervention in the mid-second century, was instigated and carried out between indigenous polities. During the first two centuries of indigenous urbanisation in the area around Martigues, large centres rise, collapse and are replaced, in the cyclical nature described by Earle as indicating powers with military, but not economic or ideological, authority. The small, compact settlements which, in the fourth century BC, replace the sprawling, but ultimately unstable, early centres, recall periods of 'recession' described by both Carneiro (1970) and Earle (1997). The early-third century saw the emergence of an iconography of expansion and the apparent centralisation of surplus and trade production at centres in this region, and after a century in which small-scale warfare may have been rife, this (proposed) conflict accelerated towards the concentration of aggression in around 200 BC. It seems unlikely that the outbreak of warfare in around 200 BC and the subsequent rise in socio-political complexity are unrelated, reinforcing the hypothesis that the earlier violence in the region was internally driven and politically motivated.

External Warfare

The consequences of a permanent, state-level presence (Massalia) in a territory occupied by tribal or chiefdom-level societies may have been many and varied. The most obvious outcome may have been straightforward colonial conquest for sites which the Phocaeans found desirable or the destruction of certain sites to close down competing indigenous trade centres. Another result of such a presence may have been the intensification of internecine conflict, as indigenous centres vied for a place in these new trading networks. This last may have accelerated existing competition, or introduced new players into the arena, as, for example, territories previously considered unfertile, became productive in the production of vines or olives for trade.

State-level (essentially Greek) activity in Provence during the Iron Age probably incited both inter-ethnic and internecine warfare. It would seem unlikely that the foundation of permanent Greek colonies and the associated imposition of different forms of occupation and exploitation of the landscape would not result in some form of conflict with the indigenous populations. The nature of the Phocaean colonisation, however, particularly after around 450 BC, when Greek attention turned from the Rhône corridor to the Mediterranean littoral, demanded good, or at least stable relationships with their Gaulish neighbours. On the other hand, the Massaliote chora is generally understood to be small, and as a primarily mercantile presence, the relationships of

exchange, particularly with those controlling food-producing resources, must have sparked competition between Gaulish communities.

The Phocaean colonisation may have created various kinds of civilised-tribal zone violence (inter-ethnic and internecine) all or any of which may have guided the path of indigenous political centralisation and the emergence of the Saluvian centres. Contrary to traditional interpretations much of this conflict appears to have taken place in contexts of competition and expansion within the indigenous milieu. Nevertheless, the changes in indigenous strategies of economic or political organisation owe much to the Phocaean presence.

CONCLUSION

9.1 Introduction

The Iron Age of southern F[...] [...] anglophone research. The few accounts published in English over the years do, however, seem to suggest a society in which warfare figured strongly. Settlements such as Entremont and Saint-Blaise appeared to demonstrate that concentrations of the population had aggregated behind impressive fortifications. The warrior statues such as those from Entremont and Roquepertuse suggested a culture imbued by war. Perhaps most compellingly of all, the cranial remains from Entremont, Roquepertuse and Glanum seemed to support the accounts of Poseidonius and Polybius, which described the Gauls as headhunters in war. The foregoing analyses have examined the documentary, settlement, iconographic and osteological evidence for warfare in this region, each within its chronological context and in tandem with one another. The study has also tested some more exploratory forms of evidence based on anthropological studies of warfare. The reliability, limitations and potential of these forms of evidence to identify warfare have been discussed in the last chapter and are assessed below.

The study has demonstrated some of the ways that warfare might be integrated into wider archaeological interpretations. The case-study of the southern French Iron Age has shown that, far from being an episodic series of events, warfare is actually a powerful causal agent with far-reaching influences on social and political structures.

9.2 Evaluation of the Evidence

The Documentary Evidence

Chapter four examined the contemporary or near-contemporary accounts of warfare and warlike behaviour among the Gauls, Greeks and Romans in southern France including both historical accounts of conflict and ethnographic descriptions of warlike practices, particularly predatory headhunting and the 'Celtic cult of the head'. These literary accounts correspond best with the later Roman campaigns, while earlier accounts of warfare were limited to vague references to 'unrest'. In general, only the events which brought Greeks or Romans into direct conflict with the Gauls were recorded. This is potentially very misleading, as the direct archaeological evidence for an episode of widespread aggression around 200 BC shows.

The 'war-trophy' interpretation of cranial remains from southern French oppida has been largely based on the accounts left by Poseidonius, who is believed to have visited Massalia around 90 BC. This places the historian around 15 km from La Cloche, where heads were displayed in disfiguring armatures above the entrance-gate in the mid first century BC. Poseidonius' account within its very particular and volatile historical and political context, however, cannot be said to relate to earlier depositions and displays of cranial remains in the region.

The Settlement Evidence

Direct Evidence for Aggression

Direct evidence for aggression has provided the most unequivocal evidence for warfare in this study and, to some extent, has acted as a touchstone for other, more exploratory, forms of evidence for warfare. In establishing patterns of warfare for this period, direct evidence of warfare has been useful in many ways, particularly in presenting for the first time the extent of warfare at the beginning of the second century BC. However, the lack of evidence for unenclosed settlement and burial throughout the Iron Age has created certain limitations. The direct evidence for assault at the sites in this study very likely reflects the culmination of periods of conflict carried out among the 'invisible' populations dispersed among territories of which these enclosed settlements may merely have been the refuges and/or political centres.

Evidence for Intention to Defend Settlements

The evidence for defensive intention corresponded surprisingly well with that for direct evidence for aggression suggesting that the former may indeed be legitimate evidence for perceived threat, and also that these two forms of evidence may be useful as an indication of large-scale and/or externally directed warfare.

Socialisation for Mistrust

The Ember and Ember (1992) cross-cultural study quantified the effects of recurrent warfare on small-scale societies. Their findings suggested that warfare was precipitated by fear and caused a breakdown of trust relationships within communities outside the immediate family. This study looked for artefactual evidence for mistrust (securing personal belongings and superstition) and structural evidence for the fragmentation of the community (small households, restrictions of access around settlements and an avoidance of communal space).

The evidence for securing personal goods and superstition raised problems of survival. Keys, locks and caskets etc. are either plundered by those assaulting a settlement, or retrieved by occupants returning to salvage their belongings. Many of the most extensively excavated sites, however, yielded evidence for caches of coins, or valuables in jars or hidden in house walls. The attempt to identify evidence for superstition was disappointing as anthropological studies, both anecdotal (e.g. Ferguson 1992; Zikmund and Hanzelka 1963) and analytical (Marwick 1970; Ember and Ember 1991) suggest that there is a correlation between superstition and warfare. It may be significant that the two slave burials/sacrifices dating to the second and first centuries BC correlate with the most intense period of warfare.

Communal space as an indicator of fragmentation was problematic on two main counts. Firstly, smaller floor-space may increase the need for work or storage areas outside the household. This could mean that, rather than fragmenting the community into family groups, very small domestic space may in fact reflect greater interaction and trust between members of the community. Secondly, the potential relationship between variations in the degree of communal space and warfare is complicated by the likelihood that more complex settlements would be expected to have public buildings or meeting places regardless of recurrent warfare (as at Entremont in the late-second/early-first century). This last problem may indicate that, like other forms of evidence for mistrust, the absence of shared space may be more likely to correlate with warfare in less-complex societies.

An unlooked for outcome of the analysis of shared space was the possibility that utilitarian shared space within settlements may indicate the beginnings of overarching economic control as outlined in Earle's (1997) theory of political economy. Viewed in this way, changes in settlement layout which include reduced domestic space and the appearance of shared work and storage spaces may have the potential to identify economic centralisation and recurrent, small-scale warfare: a situation which can result in an increase in socio-political complexity.

The attempt to plot the ease (or otherwise) of movement around settlements through access analysis was limited by the partial site plans available for most settlements of the southern French Iron Age. Where phases of one settlement (notably L'Île) were available, the restriction of access in the rebuilding after phases of destruction was interesting. The occurrence of narrowed or blocked streets (e.g. Le Baou-Roux) prior to an assault at some settlements also suggested that restrictions of access around sites was related to periods of conflict. Analysis of relative house sizes has identified a period of greatest fragmentation of communities in the third to early-second centuries BC. Comparison of the very small houses of the third century to other strands of evidence in this study has suggested that small households may potentially indicate periods of internal warfare among small-scale, dispersed populations, which conventional archaeological approaches are unable to identify.

While conflict between complex societies might be reflected in the state-sanctioned manifestations of warfare (e.g. more powerful armies, fortifications etc.); within the small-scale societies of the Embers' (1992) study, manifestations of 'mistrust' (a strong indicator of recurrent small-scale warfare) were seen on a domestic level; in particular, in an aversion to communality. Prior to the period of increased centralisation in the mid-second century, the period of greatest fragmentation in the third century corresponds with the least evidence for communality (as well as with militaristic iconography, suggesting an increasing enculturation of warrior values)

(see table 8.1). This suggests that fragmentation may be a useful indicator of recurrent warfare among more dispersed, small-scale societies.

The Iconographic Evidence

Within the southern French Iron Age, iconographic themes seem to reflect the transformations of chiefs and the societies they commanded. The dynamic war-chiefs of the seventh-century stelae at Glanum and sixth-century sculpture at Lattes were replaced by ritual leaders in the fifth century. It was demonstrated in the previous chapter that many of the sites which yielded such remains went on to become powerful centres (e.g. Entremont, Pierredon, and Glanum). This demonstrates the potential of iconographic evidence to identify the consolidation of ideological power in emerging chiefdoms. Perhaps the most useful application of the iconographic evidence in the context of this study, however, was the way in which the expansionist iconography of the early-third century (e.g. Entremont and Roquepertuse) appeared to support the hypothesis of a period of small-scale internecine, but expansionist, warfare.

The Osteological Evidence

Osteological remains comprised the most limited strand of evidence available in the study. Most consist of crania which are so heavily ritualised in their treatment and deposition that interpretations relating to warfare or violence are difficult to substantiate. Only the skull caches from Glanum appear to represent warriors killed in battle. The various cranial remains from Roquepertuse appear to have been displayed in a context of veneration, (although there are ethnographic examples of heads taken in violence which are subsequently transformed into revered and valued ritual objects). In the context of this study, the osteological remains are most usefully considered, not as a potential indicator of violence, but as a form of iconography.

9.3 Summary

Each strand of evidence has its own limitations, and comparisons between them often produced seemingly contradictory results. These apparent contradictions, however, may suggest that specific kinds of evidence reflect specific kinds of warfare. For example, fragmentation, along with economic and ideological indicators of increasingly centralised leadership, may have implications for recognising small-scale patterns of warfare which more conventional forms of evidence fail to identify. Despite the apparent discrepancies it appears clear that, in order to identify the full scale and nature of warfare, all the strands of evidence must be viewed together.

Finally, the interpretation of warfare in the southern French Iron Age discussed in the previous chapter has demonstrated how, if treated as a form of social-interaction rather than a temporary break-down of social 'norms', warfare may be integrated into archaeological interpretations of social and political transformations.

Some of the sites included in this study, such as Entremont and Roquepertuse, elicit enough general interest to generate long-term excavation and analysis and regular, detailed, published results of these investigations. More often sites are either wholly unpublished or stored as typescripts in university libraries, inaccessible to the public. Much of the information required to undertake this analysis was compiled through examination of the annual Bilan Scientifique reports stored at the regional Service Regionale d'Archéologie, and the regular and topical publication of notable finds and results in the *Documents d'Archéologie Méridionale*. The contemporary French interpretations of these finds were available through various syntheses of material published recently by, for example, Dominique Garcia (2004), Chausserie-Laprée (2005) and Patrice Arcelin and Xavier Delestre (2005). The limited access to excavation reports for the full excavated area has meant that house sizes have often been taken from averages taken from more general publications (when, for present purposes, a mean would have been preferable). The sources for house sizes is listed below.

Relative Household Sizes throughout the Iron Age

The 'small' household range = $22.86 - 40m2$ corresponded with the larger households of the later 2nd century. For this reason, relative household sizes were used to determine changes throughout the period. The average household size in chapter five = range of $16 - 19m2$. Chapter eight (Table 8.1) shows the smallest, smaller than average, average and larger than average household sizes. For sites see Table 5.3.

Sixth Century

$20 + 22 / 2 = 21m2$
Fifth Century

$20 + 16 + 13 + 24 / 4 = 18m2$
Fourth Century

$8 + 17 + 11 + / 3 = 12m2$
Third Century

$8 + 10 + 11 + 20 + 18 / 5 = 11m2$

Early Second Century

$13 + 11 + 25 + 10 / 4 = 15m2$
Late Second Century

$25 + 13 + 49 / 3 = 29m2$
Early First Century

$40 + 49 / = 44m2$

Determining Average Household Sizes

Tamaris

Averages and means calculated by the author from simplified site plans of partial excavations by Duval (from Duval 2005, 134 – 37).

Tamaris Zone 2	Tamaris Zone 3
3.75 x 5 = 18.75m2	5 x 3.75 = 18.75m2
4.45 x 3.75 = 16.75m2	3.75 x 5 = 18.75m2
5.63 x 3.75 = 21.11m2	3.75 x 5 = 18.75m2
6.25 x 5 = 31.25m2	5 x 3.75 = 18.75m2
5 x 5 = 25m2	5 x 3.75 = 18.75m2
3.75 x 5 = 18.75m2	3.75 x 4.4 = 16.75
7.5 x 3.13 = 23.5	4.45 x 3.75 = 16.75m2

7.5 x 3.75 = 28.13m2	4.45 x 3.75 = 16.75m2
3.75 x 3.75 = 14.06m2	5. x 2.5 = 12.5m2
5.63 x 3.75 = 21.11m2	6.25 x 2.5 = 15.6m2
6.88 x 3.75 = 25.80m2	3.75 x 3.75 = 14m2
	3.75 x 6.25 = 23.5m2
	5 x 5 = 25m2

Tamaris Zone 2	*Tamaris Zone 3*
Mean = 21m2	Mean = 18.75m2
More than half the houses = 20m2	Five out of the 13 houses = 18.75m2
Average house size = 25.43m2	Average house size = 18m2

Saint-Pierre

Chausserie-Laprée (2005, 130-31) based on his own excavations between 1998 – 2001 taking in the entire undamaged surface area

L'Arquet

From Lagrand (unpublished report) cited in Chausserie-Laprée (1990, 56) from partial excavation of incomplete central islet

L'Île

Chausserie-Laprée *et al.* (1987, 43 - 48) from his excavations with Nin between 1978 -1998 taking in the entire surface area

Notre-Dame-du-Pitié

From Gantès (1990, 74) from his own excavations in the late 1970s-early 1980s taking in the entire (limited) undamaged surface area

Roquepertuse

From Boissinot & Gantès (2000, 257 and 263) from their own excavations between 1994 – 1999 taking in all the known (limited) undamaged area

Le Baou-Roux

Boissinot (1990, 93) from his own partial excavations of the (limited) undamaged surface area

Entremont

Arcelin *et al.* (1990, 104) and Arcelin (1993, 72) from his own partial excavations from 1983

La Cloche

Chabot (2004, 69) from his own excavations between 1974 – 2000 taking in almost the entire surface area

APPENDIX 2: DEFENSIVE INTENTION FROM THE SIZE AND STRATEGIC POSITIONING OF THE FORTIFICATIONS

Rampart dimensions	Topographical defense	Total
1-2m (small) = 1	*Limited natural defense = 1*	*1 - 2 = little intention*
2-3m (medium) = 2	*Poorly used natural defense =2*	*3 - 5 = average intension*
3-4m (large) = 3	*Utilised natural defense = 3*	*6 - 8 = serious intension*
4m+ (extra-large) = 4	*Utilised natural defense + exposure/ hardship = 4*	

Saint-Blaise (rampart = 4 / topographical defence = 3)
Rampart: c5m thick = 4
Hilltop site between the Étangs de Citis and Lavalduc. The rampart was constructed on the break of slope, with outworks and ditches at the most vulnerable points around the settlement = 3
(Sources: Arcelin *et al.* 1983; Bouloumié & Tréziny in (eds) Arcelin & Dedet 1985).

Tamaris (rampart = 4 / strategic defence = 4)
Double rampart: c3m thick and 2.5m thick with 1m interior reinforcement = 4
An *éperon barré* – protected by cliffs, the outer rampart was constructed at the break of slope, towers and a huge strategic gateway protected the only accessible point. The excavators (Lagrand and later Duval) believe that the double rampart (cutting the site into two areas) served a social function, separating inhabitants of different status or involved in different activities. While this may be the case, the baffle-gate in the southern rampart acts as an extremely efficient protection to the only access point to the southern part of the settlement: a small cove, accessible only by sea. This may suggest that the southern zone of Tamaris acted as a refuge for the community.
Exposed to erosion and violent winds = 4

Castellan (rampart = 1 / strategic defence = 3)
Rampart = 1m
Hilltop site, rampart on break of slope = 3

L'Île (rampart = 2 / strategic defence =4)
Rampart = 2.10m
Towers along land-facing side, outwork on northwest (landward side) only one known entrance on water-facing side.
Exposed to subsidence as artificial ground surface continued to sink throughout the use of the site = 4
(Source: Chausserie-Laprée & Nin in (eds) Arcelin & Dedet 1985)

L'Arquet (rampart = 2 / strategic defence =4)
Rampart: possibly c. 3m thick = 4
Éperon barré – protected by cliffs
Exposed to erosion and violent winds = 4
(Source: Chausserie-Laprée 2005, Chapter 3 *'Villages Gauloise de l'Ouest de L'Étang de Berre: Les Fortifications'*)

Saint-Pierre (rampart = 1/ strategic defence =1)
Rampart 1 – 2m thick = 1
On small hill in shallow valley, towers appear primarily for observation = 1
(Source: Chausserie-Laprée 2005, Chapter 3 *'Villages Gauloise de l'Ouest de L'Étang de Berre: Les Fortifications'*)

Notre-Dame-du-Pitié (rampart = 4/ strategic defence = 2)
Rampart: 2 – 5.5m thick = 4
Upland plateau, outworks and towers at gate (most vulnerable point), however, this gate is placed on easiest approach and natural defences are not utilised (widest ramparts and largest towers along cliff-edge) suggesting display rather than defence was the priority = 2

Le Baou-Roux (rampart = 4 / strategic defence = 3)
Rampart 2.5 – 5m wide = 4
On high plateau, ramparts c.2.5m but constructed in a double, parallel line (5m) to protect the narrow approach = 3
(Source: Boissinot in (eds) Arcelin & Dedet 1985)

Buffe-Arnaud (rampart = 4 / strategic defence = 3)
Ramparts: 5m
Hilltop site protected by large gate tower
(Source: Arcelin & Garcia 1995)

Constantine (rampart = 3/ strategic defence = 3)
Rampart: variable but greater than 3m thick.
Hilltop site. Two quadrangular towers 3m wide and projecting 2.5m from the curtain. Around 125 BC (just prior to the destruction) round towers were built around the existing quadrangular ones, adding 0.5m to their mass, and affording better protection against assault.
(Sources: Leveau 1990, 112-17; Verdin *et al.* 1991, 110 (*Bilan Sci*)).

Entremont (rampart = 3/ strategic defence = 3) strategic

Ville Haute
Rampart: 1.5m thick = 1

Ville Basse
Rampart: 3.5m thick = 3
Hilltop site protected by cliffs, with fortifications along the only approachable side to the north.

La Cloche (rampart = 1/ strategic defence = 3)
Rampart 1.3m thick = 1
On a high hill protected by cliffs. Chabot (1985) describes the rampart as a surrounding wall rather than a fortification, and does not believe defence to be a priority. Nevertheless, the single approach road is controlled in the plain below the hill, and narrow (less than a metre) as it rounds the top of the hill, with the walls of the settlement on one side and a sheer drop on the other as it enters the settlement.
(Source: Chabot in (eds) Arcelin and Dedet 1985)

Fortifications in the Southern French Iron Age

The potential defensive function of ramparts in Mediterranean France has already been discussed in some detail in a volume edited by Arcelin and Dedet (1985). In this publication, *Les Enceintes Protohistoriques de Gaule Meridionale*, excavators of various sites assess the extent to which the ramparts enclosing indigenous settlements in the Iron Age of southern France were defensive, for display or simply to delineate the settlement area. The evaluation of defensive intention above follows to some extent the opinions of the excavators regarding rampart function and placement. Where these opinions are rejected or appended (e.g. Tamaris or La Cloche) is indicated. The suggestion that certain hardships may have been endured in the interests of defence can also be attributed to the present study.

BIBLIOGRAPHY

Aldhouse-Green, M. J. 1989. *Symbol and Image in Celtic Religious Art*. London: Routledge.

Aldhouse-Green, M. J. 2001. *Dying for the Gods: Human Sacrifice in Iron Age and Roman Europe*. Stroud: Tempus.

Alexander, R. D. 1979. 'Evolution and culture', in Chagnon, N. A. & Irons, W. (eds) *Evolutionary Biology and Human Social Behavior: An Anthropological Perspective*, 59-78. Massachusetts: Duxbury Press.

Anati, E. 1994. *Valcamonica Rock Art: A New History for Europe*. Capo di Ponte: Edizioni del Centro.

André, L. & Charrière, J. L. 1998. 'Histoire des recherches à l'oppidum d'Entremont, Aix-en-Provence', *Documents d'Archéologie Méridionale* 21, 11-20.

Andrieu-Ponel, V., Ponel, P., Bruneton, H., Leveau, P. & de Beaulieu, J. L. 2000a. 'Palaeoenvironments and cultural landscapes of the last 2000 years reconstructed from pollen and coleopteran records in the Lower Rhône Valley, southern France', *The Holocene* 10, 3, 341-55.

Andrieu-Ponel, V., Ponel, P., Jull, A. J. T., de Beaulieu, J.-L., Bruneton, H. & Leveau, P. 2000b. 'Towards a reconstruction of the Holocene vegetation history of Lower Provence: two new pollen profiles from Marais des Baux', Vegetation, History and Archaeobotany 9, 71-84.

Arcelin, P. 1991. 'Céramiques campaniennes et dérivéés régionales tardives de Glanum (Saint-Rémy-de-Provence, B. du Rh.) Questions culturelles et chronologiques', *Documents d'Archéologie Méridionale* 14, 205-38.

Arcelin, P. 1992. 'Salles hypostyles, portiques et espaces cultuels d'Entremont et de Saint-Blaise (B. du Rh.)', *Documents d'Archéologie Méridionale* 15, 13-27.

Arcelin, P. 1993. 'L'habitat d'Entremont: urbanisme et modes architecturaux', in Coutagne, D. (ed.), *Archéologie d'Entremont au Musée Granet*, 57-100. Aix-en-Provence: Association des Amis de Musée Granet.

Arcelin, P. 2000. 'Honorer les dieux et glorifier ses héros. Quelques pratiques cultuelles de la Provence gauloise', in Chausserie-Laprée (ed.) *Le Temps des Gaulois en Provence*, 92-103. Marseille: Musee Ziem.

Arcelin, P. 2004. 'Entremont et la sculpture du second âge du fer en Provence', *Documents d'Archéologie Méridionale* 27, 71-84.

Arcelin, P. 2005. 'L'aristocratie celtique et ses representations', in Delestre, X. (ed.) *15 Ans d'Archéologie en Provence-Alpes-Côtes d'Azur*, 160-67. Aix-en-Provence: Edisud.

Arcelin, P., Congès, G. & Willaume, M. 1990. 'Entremont', in Arcelin, P. & Treziny, H. (eds) *Les Habitats Indigènes des Environs de Marseille Grecque: Voyage en Massalie 100 Ans d'Archéologie en Gaule du Sud,* 100-111. Marseille: Musées de Marseile/Edisud.

Arcelin, P., Dedet, B. & Schwaller, M. 1992. 'Espaces public, espaces religieux protohistoriques en Gaule méridionale', *Documents d'Archéologie Méridionale* 15, 181-248.

Arcelin, P. & Rapin, A. 2002. 'Images de l'aristocratie du second âge du fer en Gaule méditerranéenne. Autour de la statuaire d'Entremont', in Guichard, V. & Perrin, F. (eds) *L'aristocratie celte à la fin de l'âge du fer (IIe s. av. J.-C. – Ie s. ap. J.-C.), Actes de la table ronde internationale du CAE du Mont Beuvray*, 1999, 29-66. Glux-en-Glenne: Bibracte 5.

Arcelin, P. & Rapin, A. 2003. 'Considérations nouvelles sur l'iconographie anthropomorphe de l'âge du fer, en Gaule méditerranéenne', in Büchsenschutz, O., Bulard, A., Chardenoux, M.-B. & Ginoux, N. (eds) *Décors, Images et Signes de l'Âge du Fer Européen*, 183-221. Tours: Association Française pour l'Étude de l'Âge du Fer et Fédération pour l'Édition de la Revue Archéologique du Centre de la France.

Arcelin, P., Pradelle, C., Rigoir, J. & Rigoir, Y. 1983. 'Note sur des structures primitives de l'habitat protohistorique de Saint-Blaise (Saint-Mitre-les-Remparts, B.-du-Rh.)', *Documents d'Archéologie Méridionale* 6, 138-43.

Armit, I. 1997. *Celtic Scotland*. London: Batsford.

Armit, I. 2006. 'Inside Kurtz's compound: headhunting and the human body in prehistoric Europe', in Bonogofsky, M. (ed.) *Skull Collection, Modification and Decoration*, 1 – 14. Oxford: British Archaeological Reports, International Series 1539.

Armit, I. 2011. 'Headhunting and social power in Iron Age Europe', in T. Moore & Armada, X.L. (eds) *Atlantic Europe in the First Millennium BC: Crossing the Divide*. Oxford: Oxford University Press.

Arnott, G. 'Athenaeus and the epitome: texts, manuscripts and early editions', in D. Braund & J. Wilkins (eds) *Athenaeus and His World: Reading Greek Culture in the Roman Empire*, 41-52 and 542f. Exeter: University of Exeter Press.

Adouze, F. & Büchsenschutz, O. 1989. *Towns, Villages and Countryside of Celtic Europe*. London: Batsford.

Avery, M. 1986. 'Stoning and fire at hillfort entrances of southern Britain', *World Archaeology* 18, 2, 216 – 30.

Barbet, A. 1991. 'Roquepertuse et la polychromie en Gaule méridionale à l'époque préromaine', *Documents d'Archéologie Méridionale* 14, 53 – 81.

Bar-Yosef, O. & Meadows, R. 1995. 'The origins of agriculture in the Near East', in Price, T. D. & Gebauer, A. G. (eds) *Last Hunters – First Farmers: New Perspectives on the Prehistoric Transition to Agriculture*, 39 – 94. Santa Fe: School for American Research.

Barruol, G., Gibert, U. & Rancoule, G. 1961. 'Le défunt héroïsé de Bouriège', *Revue d'Études Ligures*, XXVII, 45 – 60.

Beckerman, S. & Lizarralde, R. 1995. 'State-tribal warfare and male-biased casualties among the Barí', *Current Anthropology* 36, 497 – 500.

Bennett Ross, J. 1984. 'Effects of contact on revenge hostilities among the Achuarä Jívaro', in R. B. Ferguson (ed.) *Warfare, Culture and Environment*, 83 – 110. London: Academic Press Inc.

Benoît, F. 1955. 'Le sanctuaire aux 'esprits' d'Entremont', *Cahiers Ligures de Préhistoire et d'Archéologie* 4, 38 – 69.

Benoît, F. 1964. 'Les 'têtes sans bouche' d'Entremont', *Cahiers Ligures de Préhistoire et d'Archéologie, Bordighera* VIII, 68 – 81.

Benoît, F. 1969. *L'art primitif méditerranéen de la vallée du Rhône*. Aix-en-Provence: Publications des Annales de la Faculté des lettres d' Aix-en-Provence.

Benoît, F. 1975. 'The Celtic oppidum of Entremont, Provence', in Bruce-Mitford, R. (ed.) *Recent Archaeological Excavations in Europe*, 227 – 59. London: Routledge.

Bessac, J.-C. 1991. 'Roquepertuse. Techniques du travail de la pierre, repères chronologiques', *Documents d'Archéologie Méridionale* 14, 43 – 51.

Bessac, J.-C. & Chausserie-Laprée, J. 1992. 'Documents de la vie spirituelle et publique des habitats de Saint-Pierre et de L'Île à Martigues (B. Du Rh.)', *Documents d'Archéologie Méridionale* 15, 134 – 57.

Biolsi, T. 1984. 'Ecological and cultural factors in Plains Indian warfare', in Ferguson R. B. (ed.) *Warfare, Culture and Environment*, 141 – 68. London: Academic Press Inc.

Blanton, R. E. 1994. *Houses and Households: A Comparative Study (Interdisciplinary Contributions to Archaeology)*. London: Plenum Press.

Blick, J. 1988. 'Genocidal warfare in tribal societies as a result of European-induced culture conflicts', *Man* 23, 654 – 70.

Bonhage-Freund, M. T. & Kurland, J. A. 1994. 'Tit-for-tat among the Iroquois: a game theoretic perspective on inter-tribal political organisation', *Journal of Anthropological Archaeology* 13, 3, 278 – 305.

Bouby, L. & Marinval, P. 2001. 'La vigne et les debuts de la viticulture en France: apports de l'archeobotanique', *Gallia* 58, 13 – 28.

Bouloumié, B. 1990. 'Saint-Blaise', in Arcelin, P. & Treziny, H. (eds) *Les Habitats Indigènes des Environs de Marseille Grecque: Voyage en Massalie 100 Ans d'Archéologie en Gaule du Sud*, 33 – 41. Marseille: Musées de Marseille/Édisud.

Boissinot, P. 1994. 'Velaux, Roquepertuse', *Bilan Scientifique*, Aix-en-Provence, (D. R. A. C. / S. R. A.) 1994, 164 – 66.

Boissinot, P. 1998. 'La reintérpretation du sanctuaire de Roquepertuse', *Archéologia* 351, 42 – 45.

Boissinot, P. 1999. 'Velaux, Roquepertuse', *Bilan Scientifique*, Aix-en-Provence, (D. R. A. C. / S. R. A.) 1999, 117 – 18.

Boissinot, P. 2000. 'Velaux, Roquepertuse', *Bilan Scientifique*, Aix-en-Provence, (D. R. A. C. / S. R. A.) 2000, 137 – 38.

Boissinot, P. 2004. 'Usage et circulation des éléments lapidaries de Roquepertuse', *Documents d'Archéologie Méridionale* 27, 49 – 62.

Boissinot, P. & Gantès, L.-F. 2000. 'La chronologie de Roquepertuse. Propositions préliminaries a l'issue des campagnes 1994 – 1999', *Documents d'Archéologie Méridionale* 23, 249 – 71.

Bonefant, P.-P. & Gillaumet, J.-P. 2002. État des recherches effectuées par les auteurs sur la sculpture préromaine en Europe', *Documents d'Archéologie Méridionale* 25, 257 – 60.

Bowden, M. & McOmish, D. 1987. 'The required barrier', *Scottish Archaeological Review* 4, 76 – 84.

Bowden, M. & McOmish, D. 1989. 'Little boxes: more about hillforts', *Scottish Archaeological Review* 6, 12 – 6.

Braund, D. 2000. 'Introduction', in Braund, D. & Wilkins, J. (eds) *Athenaeus and his World: Reading Greek Culture in the Roman Empire*. Exeter: University of Exeter Press.

Bridgford, S. D. 1997. 'Mightier than the pen? (an edgewise look at the Irish Bronze Age swords)', in Carman (ed.) *Material Harm*, 95 – 115. Glasgow: Cruithne Press.

Brion, M. 1956. *Provence*. London: Nicholas Kaye Ltd.

Briscoe, J. 1971. 'Livy in the first decade', in Dorey, T. A. (ed.) *Livy*, 1 – 20. London: Routledge.

Brogan, O. 1974. 'The coming of Rome and the establishment of Roman Gaul', in Piggott, S., Daniel, G. & McBurney, C. (eds) *France Before the Romans*, 192 – 219. London: Thames and Hudson.

Brun, J.-P. 1993. 'Les huileries d'Entremont', in Coutagne, D. (ed.), *Archéologie d'Entremont au musée Granet*, 101 – 05. Aix-en-Provence: Association des Amis de Musée Granet.

Brun, J.-P. & Laubenheimer, F. 2001. 'Introduction', in *La Viticulture en Gaule*, 1 – 4. (Gallia supplement) Paris: C. N. R. S. Éditions.

Buffat, L. & Pellecuer, C. with contributions by Mauné, S. & Pomarédes, H. 2001. 'La viticulture antique en Languedoc-Roussillon', in Brun, J.-P. & Laubenheirmer (eds) *La Viticulture en Gaule*, 91 – 111. (Gallia supplement) Paris: C. N. R. S. Éditions.

Buxo-I-Capdevila, R. 1996. Evidence for vines and ancient cultivation from an urban area, Lattes (Herault), southern France. *Antiquity* 70, 393.

Byrd, B. F. 2000. 'Households in transition: Neolithic social organisation within Southwest Asia', in Kuijt, I. (ed.) *Life in Neolithic Farming Communities: Social Organisation, Identity and Differentiation*, 63 – 98. New York: Kluwer Academic / Plenum Publishers.

Carman, J. 1997. *Material Harm*. Glasgow: Cruithne Press.

Carman, J. & Harding, A. 1999. *Ancient Warfare*. Stroud: Sutton Publishing Ltd.

Carneiro, R. 1970. 'A theory of the origin of the state', *Science* 169, 733 – 38.

Carneiro, R. 1978. 'Political expansion as an expression of the principle of competitive exclusion', in Cohen, R. & Service, E. (eds) *Origins of the State: the Anthropology of Political Evolution*, 205 – 225. Philadelphia: Institute for the Study of Human Issues.

Carneiro, R. 1990. 'Chiefdom-level warfare as exemplified in Fiji and the Cauca Valley', in Haas, J. (ed.) *The Anthropology of War*, 190 – 211. Melksham: Redwood Press Ltd.

Chabot, L. 1983a. 'L'oppidum de La Cloche aux Pennes Mirabeau (Bouches-du-Rhone) (synthese de travaux effectues de 1967 à 1982)', *Revue Archéologique de Narbonnaise* 16,38 – 80.

Chabot, L. 1983b. 'L'oppidum de La Cloche aux Pennes Mirabeau (Bouches-du-Rhone)', *Revue Archéologique de Narbonnaise* 16, 39 – 40.

Chabot, L. 1990. 'La Cloche' in Arcelin, P. & Treziny, H. (eds) *Les Habitats Indigènes des Environs de Marseille Grecque: Voyage en Massalie 100 Ans d'Archéologie en Gaule du Sud*, 118 – 25. Marseille: Musées de Marseille / Édisud.

Chabot, L. 1992. 'Pennes Mirabeau, La Cloche', *Bilan Scientifique, Aix-en-Provence*, (D. R. A. C. / S. R. A.) 1992, 149 – 50.

Chabot, L. 1993. 'Pennes Mirabeau, La Cloche', *Bilan Scientifique, Aix-en-Provence*, (D. R. A. C. / S. R. A.) 1993, 126 – 28.

Chabot, L. 1994. 'Pennes Mirabeau, La Cloche', *Bilan Scientifique, Aix-en-Provence*, (D. R. A. C. / S. R. A.) 1994, 151 – 52.

Chabot, L. 1996. 'Une aire cultuelle sur l'oppidum de La Cloche aux Pennes-Mirabeau (Bouches-du-Rhone) Les enseignements de la zone sommitale', *Revue Archéologique de Narbonnaise* 29, 233 – 84.

Chabot, L. 1997. 'Pennes Mirabeau, La Cloche', *Bilan Scientifique, Aix-en-Provence*, (D. R. A. C. / S. R. A.) 1997, 86 – 87.

Chabot, L. 2000. 'Pennes Mirabeau, La Cloche', *Bilan Scientifique, Aix-en-Provence*, (D. R. A. C. / S. R. A.) 2000, 127 – 28.

Chabot, L. 2004. *L'oppidum de La Cloche (Les Pennes-Mirabeau, Bouches-du-Rhône)*. Montangnac: Editions Monique Mergoil.

Chadwick, N. 1971. *The Celts*. London: Penguin Books.

Chagnon, N. 1977. *Yąnomamö: The Fierce People*. New York: Holt, Rinehart and Winston.

Chagnon, N. 1992. *Yąnomamö: The Fierce People*. London: Harcourt Brace Jovanovich College Publishers.

Chagnon, N. 1996. *Yąnomamö*. London: Harcourt Brace Jovanovich College Publishers.

Chagnon, N., Neal, J., Weitkamp, L., Gershowitz, H. & Ayres, M. 1970. 'The influence of cultural factors on the demography and pattern of gene flow from the Makiratare to the Yanomama Indians', *American Journal of Physical Anthropology* 32, 339 – 49.

Chagnon, N. & Melacon, T. 1983. 'Epidemics in a tribal population', in Clay, J. (ed.) *The Impact of Contact: Two Yanomamo Case Studies*, 53 – 78. Cambridge: Cultural Survival and Bennington College.

Charrière, J.-L. 1980. 'Un torse préromain découvert près de l'oppidum de Constantine (commune de Lançon. B. Du Rh.)', *Documents d'Archéologie Méridionale* 3, 159 62.

Chaume, B. 2001. *Vix et son Territoire a l'Âge du Fer. Fouilles du Mont Lassois et Environnement du Site Princier*. Mergoil: Montagnac (Protohistoire Européenne).

Chaume, B. & Reinhard, W. 2003. 'Les statues de Vix: images héroïsées de l'aristocratie Hallstattienne', *Madrider Mitteilungen* 44, 249 – 67.

Chausserie-Laprée, J. 1990. 'Martigues', in Arcelin, P. & Treziny, H. (eds) *Les Habitats Indigènes des Environs de Marseille Grecque: Voyage en Massalie 100 Ans d'Archéologie en Gaule du Sud*, 52 – 71. Marseille: Musées de Marseille/Édisud.

Chausserie-Laprée, J. 1998. 'Saint Pierre les Martigues', *Bilan Scientifique, Aix-en-Provence*, (D. R. A. C. / S. R. A.) 1998, 92 – 94.

Chausserie-Laprée, J. 1999. 'Saint Pierre les Martigues', *Bilan Scientifique, Aix-en-Provence*, (D. R. A. C. / S. 1999, 101 – 02.

Chausserie-Laprée, J. 2000. 'Saint Pierre les Martigues', *Bilan Scientifique, Aix-en-Provence*, (D. R. A. C. / S. 2000, 122 – 24.

Chausserie-Laprée, J. 2005. Martigues, *Terre Gauloise: Entre Celtique et Méditerranée*. Paris: Editions Errance

Chausserie-Laprée, J., Nin, N. et Domallain, L. 1984. 'Le village protohistorique du quartier de L'Île à Martigue; du-Rh.): urbanisme et fortification', *Documents d'Archéologie Méridionale* 7, 27 – 52.

Chausserie-Laprée, J. & Nin, N. (avec les participation de Philippe Boissinot). 1987. 'Le village protohistorique d quartier de L'Île à Martigues (B.-du-Rh.): Les aménagements domestiques', *Documents d'Archéologie Méridiona* 31 – 89.

Chausserie-Laprée, J. & Nin, N. (avec les participation de Philippe Boissinot). 1990. 'Le village protohistorique du quartier de L'Île à Martigues (B.-du-Rh.): Les aménagements domestiques', *Documents d'Archéologie Méridionale* 35 – 136.

Chenorkian, R. 1988. *Les Armes Métallique dan l'Art Préhistorique de l'Occident* Méditerranéen. Paris: C. N. R. S.

Cleere, H. 2001. *Southern France*. Oxford: Oxford Archaeological Guides.

Cocco, L. 1972. *Iyëwei-Teri: Quince Años entre los Yanomamos*. Caracas: Escuela Tecnica Popular Don Bosco.

Cohen, R. 1984. 'Warfare and state formation: wars make states and states make wars', in Ferguson, R. B. (ed.) *Warfare, Culture and Environment*, 329 – 358. London: Academic Press Inc.

Cohen, R. 1985. 'Warfare and state formation', in Claessen, H. J., van de Velde, P. & Smith, M. E. (eds) *Development and Decline: The Evolution of Sociopolitical Organisation*, 276 – 89. South Hadley: Bergin and Garvey.

Coignard, R. & Coignard, O. 1991. 'L'ensemble lapidaire Roquepertuse: nouvelle approche', *Documents d'Archéologie Méridionale* 14, 27 – 42.

Collis, J. 1976. 'Town and Market in Iron Age Europe', in Cunliffe, B. & Rowley, T. (eds) *Oppida: The Beginnings of Urbanisation in Barbarian Europe*, 3 – 23. Oxford: British Archaeological Reports (supplement) Series II.

Collis, J. 1984. *Oppida: Earliest Towns North of the Alps*. Huddersfield: University of Sheffield.

Collis, J. 1995. 'The first towns', in Green, M. (ed.) *The Celtic World*, 159 – 175. London: Routledge.

Collis, J. 2003. *The Celts: Origins, Myths and Inventions*. Stroud: Tempus Publishing Ltd.

Cook, S. 1972. *Prehistoric Demography* (Addison-Wesley Module No. 16). Addison-Wesley Modular Publications.

Cunliffe, B. 1976. 'The origins of urbanisation in Britain', in Cunliffe, B. & Rowley, T. (eds) *Oppida: The Beginnings of Urbanisation in Barbarian Europe*, 135 – 61. Oxford: British Archaeological Reports (supplement) series II.

Cunliffe, B. 1978. *Rome and her Empire*. London: The Bodley Head.

Cunliffe, B. 1988. *Greeks, Romans and Barbarians: Spheres of Interaction*. London: Batsford.

Cunliffe, B. 1995. *Danebury, an Iron Age Hillfort in Hampshire Vol 6: A Hillfort Community in Perspective*. London: Council for British Archaeology Research Report 102.

Cunliffe, B. 1997. *The Ancient Celts*. Oxford: Oxford University Press.

Cunliffe, B. 2001. *Facing the Ocean: the Atlantic and its People 8000 BC – AD 1500*. Oxford: Oxford University Press.

Cunliffe, B. 2002. *The Extraordinary Voyage of Pytheas the Greek*. New York: N. Y. Walker & Co.

Dalton, G. 1977. 'Aboriginal economies in stateless societies', in Earle, T. K. & Ericson, J. E. (eds) *Exchange Systems in Prehistory*, 191 – 212. London: Academic Press.

Damotte, L. 2003. 'Mobilier céramique et faciès culturel de l'habitat gaulois de L'Île de Martigues', *Documents d'Archéologie Méridionale* 26, 171 – 234.

Dedet, B. & Schwaller, M. 1990. Pratiques cultuelles et funeraires en milieu domestique sur les oppidums languedociens. *Documents d'Archéologie Méridionale* 13, 137 – 61.

Deflem, M. 1999. 'Warfare, political leadership and state formation: the case of the Zulu kingdom 1808 – 1879', *Ethnology* 38, 371 – 91.

Delamare, F. & Guineau, B. 1991. Roquepertuse: analyse physio-chimique des couches picturales. *Documents d'Archéologie Méridionale* 14, 83 – 86.

Dietler, M. 1995. 'Early "Celtic" socio-political relations: ideological representation and social competition in dynamic comparative perspective', in Arnold, B. & Gibson, D. B. (eds) *Celtic Chiefdom, Celtic State*, 64 – 71. Cambridge: Cambridge University Press.

Dietler, M. 1997. 'The Iron Age in Mediterranean France: colonial encounters, entanglements and transformations', *Journal of World Prehistory*, 11, 3, 269 – 339.

Dietler, M. & Py, M. 2003. 'The Warrior of Lattes: an Iron Age statue discovered in Mediterranean France', *Antiquity* 77, 780 – 95.

Duday, H. 1995. 'Études des restes humains trouvés dans les décombres de la porte incendiée', Screens 32 – 8 in CD-Rom accompanying Garcia, D. & Bernard, L. 1995. 'Un témoignage de la chute de la Confederation Salyenne? L'oppidum de Buffe Arnaud (Saint-Martin-de-Brômes, Alpes-de-Haute-Provence)', *Documents d'Archéologie Méridionale* 18, 113 – 42.

Dueck, D. 2000. *Strabo of Amasia: A Greek Man of Letters in Augustan Rome*. London: Routledge.

Dufraigne, J.-J. 1995. 'Entremont', *Bilan Scientifique, Aix-en-Provence*, (D. R. A. C. / S. R. A.) 1995, 137 – 38.

Dufraigne, J.-J. 1999. 'Entremont', *Bilan Scientifique, Aix-en-Provence*, (D. R. A. C. / S. R. A.) 1999, 67 – 69.

Duval, S. 1999. 'Tamaris', *Bilan Scientifique, Aix-en-Provence*, (D. R. A. C. / S. R. A.) 1999, 104 – 06.

Duval, S. 2005. 'Les Maisons de Tamaris au VI siecle', in Chausserie-Laprée (ed.) *Martigues, Terre Gauloise: Entre Celtique et Méditerranee*, 134 – 37. Paris: Editions Errance.

Earle, T. 1997. *How Chiefs Come to Power: The Political Economy in Prehistory*. Stanford: Stanford University Press.

Ebel, C. 1976. *Transalpine Gaul: The Emergence of a Roman Province. Studies of the Dutch Archaeological and Historical Society, v. 4.* Leiden: E. J. Brill.

Eisenberg, L. 1978. 'The "human" nature of human nature', in Caplan, A. L. (ed.) *The Sociobiology Debate: Readings on Ethical and Scientific Issues*, 163 – 80. London: Harper & Row.

Ember, C. R. & Ember, M. 1992. 'Resource unpredictability, mistrust and war: a cross-cultural study', *Journal of Conflict Resolution* 36, 242 – 62.

Ferguson, R. B. 1984. 'Introduction: studying war', in Ferguson, R. B. (ed.) *Warfare, Culture and Environment*, 1 – 61. London: Academic Press Inc.

Ferguson, R. B. 1984. 'A savage encounter: western contact and the Yanomami war complex', in Ferguson, B. & Whitehead, N. (eds) *War in the Tribal Zone: Expanding States and Indigenous Warfare*, 199 – 228. Santa Fe: School of American Research Press.

Ferguson, R. B. 1990. 'Explaining war', in Haas, J. (ed.) *The Anthropology of War*, 26 – 55. Melksham: Redwood Press Limited.

Ferguson, R. B. 2001. 'Materialist, cultural and biological theories on why Yanomami make war', *Anthropological Theory* 1, 1, 99 – 116.

Ferguson, R. B. & Whitehead, N. 1984. 'The violent edge of empire', in Ferguson, R. B. & Whitehead, N. (eds) *War in the Tribal Zone: Expanding States and Indigenous Warfare*, 1 – 30. Santa Fe: School of American Research Press.

Flannery, K. V. 1972. 'The origin of the village as a settlement type in Mesoamerica and the Near East: a comparative study', in Ucko, P., Tringham, R. & Dimbleby, G. (eds) *Man: Settlement and Urbanisation*, 23 – 53. London: Duckworth.

Foster, S. M. 1989. 'Analysis of spatial patterns in buildings (access analysis) as an insight into social structure: examples from the Scottish Atlantic Iron Age', *Antiquity* 63, 40 – 50.

Frankenstein, S. & Rowlands, M. 1978. 'The internal structure and regional context of Early Iron Age society in south-western Germany', *Bulletin of the Institute of Archaeology* 15, 73 – 112.

Freeman, P. 2002) *War, Women and Druids: Eyewitness Reports and Early Accounts of the Ancient Celts*. Austin: University of Texas Press.

Fitzhugh, W. 1985. *Cultures in Contact: The European Impact on Native Cultural Institutions in Eastern North America, AD 1000 – 1800.* Washington: Smithsonian Institution Press.

Fulford, M. G. 1985. 'Roman material in barbarian society *c.* 200 BC – AD 400', in Champion, T. C. & Megaw, J. V. S. (eds) *Settlement and Society*, 91 – 108. Cambridge: Leicester University Press.

Gallet de Santerre, H. 1980. *Enserune, les Silos de la Terrasse*. Paris: Centre National de la Recherche Scientifiques.

Garcia, D. 2000. 'The process of urbanisation in southern Gaul during the Early Iron Age', *Bibracte* 4, 49 – 60.

Garcia, D. 2004. *La Celtique Méditerranéenne: Habitats et Sociétés en Languedoc et Provence VIIIᵉ – IIᵉ siècles av. J.-C.* Paris: Editions Errance.

Garcia, D. 2005a. 'Urbanisation and special organisation in southern France and northeastern Spain during the Iron Age', *Proceedings of the British Academy* 126, 169 – 86.

Garcia, D. 2005b. 'Villes et Campagnes', in Delestre, X. (ed.) *15 Ans d'Archéologie en Provence-Alpes-Côte-d'Azur*, 81 – 97. Aix-en-Provence: Edisud.

Garcia, D. & Bernard, L. 1995. 'Un témoignage de la chute de la confederation Salyenne? L'oppidum de Buffe Arnaud (Saint-Martin-de-Brômes, Alpes-de-Haute-Provence)', *Documents d'Archéologie Méridionale* 18, 113 – 42.

Gantès, F.-L. 1979a. 'Note sur les céramiques à vernis noir trouvées sur l'oppidum de la Teste-Nègre aux Pennes Mirabeau (B.-du-Rh.)', *A. E. L.* 1, 97 – 103.

Gantès, F.-L. 1979b. 'Note sur quelques fibules et bracelets trouvées sur l'oppidum de la Teste-Nègre aux Pennes Mirabeau (B.-du-Rh.)', *Bull. Arch. Prov. 3, 38 – 46.*

Gantès, F.-L. 1990a. 'Notre-Dame-du-Pitié', in Arcelin, P. & Treziny, H. (eds) Les Habitats Indigènes des Environs de Marseille Grecque: Voyage en Massalie 100 Ans d'Archéologie en Gaule du Sud, 72 – 77. Marseille: Musées de Marseille / Edisud.

Gantès, F.-L. 1990b. 'Teste-Nègre', in Arcelin, P. & Treziny, H. (eds) Les Habitats Indigènes des Environs de Marseille Grecque: Voyage en Massalie 100 Ans d'Archéologie en Gaule du Sud, 78 – 88. Marseille: Musées de Marseille / Edisud.

Gantès, F.-L. & Lescure, B. 1992. 'Velaux, Roquepertuse', *Bilan Scientifique, Aix-en-Provence*, (D. R. A. C. / S. R. A.) 1992, 155.

Gardner, R. & Heider, K. G. 1974. *Gardens of War: Life and Death in the New Guinea Stone Age*. Harmondsworth: Penguin Books Ltd.

Gazenbeek, M. 2000. 'L'habitat rural autour du Marais des Baux: évolution de l'âge du fer à la fin de l'antiquité', *Revue Archeologique de Narbonnaise* (supplement) 31, 85 – 96.

George, K. M. 1996. 'Lyric, history and allegory, or the end of headhunting ritual in Upland Sulawesi', in (ed.) Hoskins, J. *Headhunting and the Social Imagination in Southeast Asia*, 50 – 89. California: Stanford University Press.

de Gérin-Ricard, H. 1927. *Le sanctuaire préromain de Roquepertuse a Velaux (Bouches du Rhône). Étude sur l'art gaulois avant les temps classiques.* Marseille: Société de Statistique, d'Hist. et Archéo. De Marseille.

141

de Gérin-Ricard, H. 1928. 'Le sanctuaire préromain de Roquepertuse. Fouilles de 1927', *Provincia* (supplement) 8, 53 – 60.

Gibson, T. 1990. 'Raiding, trading and tribal autonomy in insular Southeast Asia', in Haas, J. (ed.) *The Anthropology of War*, 105 – 124. Melksham: Redwood Press Limited.

Gilchrist, R. 2003. 'Introduction: towards a social archaeology of warfare', *World Archaeology* 35, 1 – 6.

Goring-Morris, N. 2000. 'The quick and the dead: the social context of Aceramic Neolithic Mortuary Practices as seen from Kfar HaHoresh', in Kjuit, I. (ed.) *Life in Neolithic Farming Communities: Social Organisation, Identity and Differentiation*, 103 – 86. New York: Kluwer Academic / Plenum.

Goudineau, C. 1983. Marseille, Rome and Gaul from the third to the first century BC, in Garnsey, P., Hopkins, K. & Whittaker, C. R. (eds) *Trade in the Ancient Economy*, 76 – 86. London: Chatto and Windus.

Gould, S. J. 1978. 'Biological potential vs. biological determinism', in Caplan, A. L. (ed.) *The Sociobiology Debate: Readings on Ethical and Scientific Issues*, 343 – 354. London: Harper & Row.

Hallpike, C. R. 1973. 'Functionalist interpretations of primitive warfare', *Man* 8, 3, 451 – 70.

Gras, M. 2004. 'Les Étrusques vus de la Gaule. Échanges maritimes et implantations', *Documents d'Archéologie Méridionale* 27, 213 – 35.

Green, M. 1995. 'The gods and the supernatural: the nature of Celtic religion', in Green, M. (ed.) *The Celtic World*, 465 – 88. London: Routledge.

Guichard, C. & Rayssiguier, G. avel la coll. de Chabot, L. 1988. 'La derniere periode d'occupation de l'oppidum de Baou de Saint-Marcel a Marseille. La ceramique d'importation et le monnayage', *Documents d'Archéologie Méridionale* 11, 71 – 96.

Guilaine, J. & Zammit, J. 2001. *Le Sentier de la Guerre: Visages de la Violence Préhistorique.* Paris: Seuil.

Guillaumet, J.-P. 2000. 'Les personages accroupis: essai de classement', in Büchsenschutz, O., Bulard, A., Chardenoux, M.-B. & Ginoux, N. (eds) *Décors, Images et Signes de l'Âge du Fer Européen*, 171 – 82. Tours: Association Française pour l'Etude de l'Âge du Fer et Fédération pour l'Édition de la Revue Archéologique du Centre de la France.

Guillaumet, J.-P. & Rapin, A. 2000. 'L'art des Gaulois du Midi', in Chausserie-Laprée (ed.) *Le Temps des Gaulois en Provence*, 79 – 83. Marseille: Musée Ziem.

Guillet, E., Lelièvre, V., Paillet, J.-L., Piskorz, M., Recolin, A. & Souq, F. 1992. 'Un monument à portique tardo-hellénistique près de la source de la Fontaine, à Nîmes (Gard), *Documents d'Archéologie Méridionale* 15, 57 – 89.

Gunawardana, R. A. L. H. 1992. 'Conquest and resistance: pre-state and state expansion in early Sri Lankan history', in Ferguson, R. B. & Whitehead, N. (eds) *War in the Tribal Zone: Expanding States and Indigenous Warfare*, 1 – 30. Santa Fe: School of American Research Press.

Haas, J. 1990. 'Warfare and the evolution of tribal polities in the prehistoric southwest', in Haas, J. (ed.) *The Anthropology of War*, 171 – 189. Melksham: Redwood Press Limited.

Hameau, P. 2005. 'La société et ses représentations: les manifestation artisiques', in Delestre, X. (ed.) *15 Ans d'Archéologie en Provence-Alpes-Côte-d'Azur*, 150 – 59. Aix-en-Provence: Edisud.

Harding, A. 1994. 'Reformation in barbarian Europe, 1300 – 600 BC', in Cunliffe, B. (ed.) *Prehistoric Europe: an Illustrated History*, 304 – 335. Oxford: Oxford University Press.

Härke, H. 1979. *Settlement Types and Settlement Patterns in the West Hallstatt Province.* Oxford: British Archaeological Reports International Series 57.

Harris, M. 1984. 'A cultural materialist theory of band and village warfare: the Yanomamö test', in Ferguson, R. B. & Whitehead, N. (eds) *War in the Tribal Zone: Expanding States and Indigenous Warfare*, 111 – 140. Santa Fe: School of American Research Press.

Harrison, S. 1993. *The Mask of War: Violence, Ritual and the Self in Melanesia.* Manchester: MUP.

Hatt, J.-J. 1970. *Celts and Gallo-Romans*. London: Barrie & Jenkins.

Herrmann, F.-R., Frey, O.-H., Bartel, A., Kreuz, A. & Rösch, M. 1997. 'Ein frühkeltischer Fürstengrabhügel am Glauberg im Wetteraukreis, Hessen. Bericht über die Forschungen 1994 – 1996', *Germania* 75, 459 – 550.

Hodson, F. R. & Rowlett, R. M. 1974. 'From 600 BC to the Roman Conquest' in Piggott, S., Daniel, G. & McBurney, C. (eds) *France Before the Romans*, 157 – 191. London: Thames and Hudson.

Hole, F. 1987. *The Archaeology of Western Iran: Settlement and Society from Prehistory to the Islamic Conquest*. Washington: Smithsonian Institution Press.

Hoskins, J. 1996. 'Introduction: headhunting as a practice and as trope', in Hoskins, J. (ed.) *Headhunting and the Social Imagination in Southeast Asia*, 1 – 49. California: Stanford University Press.

Hill, R. 2002. 'Review of "prehistoric warfare in the American Southwest" by Stephen Leblanc', *Athena Review* 2, 4.

Hulme, P. 1986. *Colonial Encounters with Caribs from Columbus to the Present Day*. Oxford: Oxford University Press.

Irons, W. 1979. 'Some statements of theory', in Chagnon, N. A. & Irons, W. (eds) *Evolutionary Biology and Human Social Behavior: An Anthropological Perspective*, 1 – 3. Massachusetts: Duxbury Press.

Jacobsthal, P. F. 1941. 'Imagery in Early Celtic Art', *Proceedings of the British Academy* 27, 301 – 20.

Jacoby, F. 1923. *Die Fragmente der Griechischen Historiker* (F Gr Hist). Berlin: Weidmann.

James, S. 1999. *The Atlantic Celts: Ancient Peoples or Modern Invention*. London: British Museum Press.

Janin, T., Taffanel, O., Taffanel, J., Boisson, H., Chardenon, N., Gardeisen, A., Hérubel, F., Marchand, G., Montécinos, A. & Rouquet, J. 2002. La nécropole protohistorique du Grand Bassin II a Mailhac, Aude (VIe – Ve s. av. n. è)', *Documents d'Archéologie Méridionale* 25, 65 – 122.

Keeley, L. 1996. *War before Civilization: The Myth of the Peaceful Savage*. Oxford: Oxford University Press.

Kjuit, I. 2001. 'Place, death and the transmission of social memory in early agricultural communities of the Near Eastern Pre-Pottery Neolithic', in Chesson, M. (ed.) *Social Memory, Identity and Death: Anthropological Perspectives on Mortuary Rituals*, 80 – 131. Arlington: American Anthropological Association.

Koch, J. T. 2000. *The Celtic Heroic Age: Literary Sources for Ancient Celtic Europe and Early Ireland and Wales*. Aberystwyth: Celtic Studies Publications.

Kristiansen, K. 1998. *Europe Before History*. Cambridge: Cambridge University Press.

Kristiansen, K. 1999. 'The emergence of warrior aristocracies in later European prehistory and their long-term history', in Carman, J. & Harding, A. (eds) *Ancient Warfare*, 175 – 90. Trowbridge: Sutton Publishing.

Lagrand, C. 1979. 'Un nouvel habitat de la periode de colonisation grecque: Saint Pierre les Martigues (Bouches-du-Rhône)', *Documents d'Archéologie Méridionale* 2, 81 – 106.

Lantier, R. 1934. Le dieu celtique de Bouray: Monuments & Mémoires. *Foundation Eugène Piot* 34, 35 – 58.

Law, R. 1992. 'Warfare on the West African slave coast, 1650 – 1850', in Ferguson, R. B. & Whitehead, N. (eds) *War in the Tribal Zone: Expanding States and Indigenous Warfare*, 103 – 26. Santa Fe: School of American Research Press.

Layton, R. & Barton, R. 2001. 'Warfare and human evolution', in Fewster, K. J. & Zvelebil, M. (eds) *Ethnoarchaeology and hunter-gatherers: pictures at an exhibition* (B.A.R. International Series 995), 13 – 42. Basingstoke: The Basingstoke Press.

LeBlanc, S. A. 1971. An addition to Naroll's suggested floor area and settlement population relationship. *American Antiquity* 36, 210 – 11.

LeBlanc, S. A. 1999. *Prehistoric Warfare in the American Southwest*. Salt Lake City: The University of Utah Press.

Lejars, T. 1994. *Gourney III, les fourreaux d'épéé: le sanctuaire de Gournay-sur-Aronde et l'armements des Celtes de La Tène moyenne*. Paris: Errance.

Lekson, S. H. 2002. 'War in the Southwest, war in the world', *American Antiquity* 67, 607 – 24.

Lenerz-de Wilde, M. 1995. 'The Celts in Spain', in Green, M. (ed.) *The Celtic World*, 533 – 51. London: Routledge.

Lescure, B. 2004. 'La statuaire de Roquepertuse et ses nouveaux indices d'interprétation a l'issue des fouilles récentes', *Documents d'Archéologie Méridionale* 27, 45 – 47.

Lescure, B. & Gantès, L.-F. 1991. 'Roquepertuse, nouvelle approche des collections', *Documents d'Archéologie Méridionale* 14, 9 – 18.

Leveau, P. 1999. 'The Integration of Archaeological, Historical and Palaeoenvironmental Data at the Regional Scale: the Vallée de Baux, Southern France', in Leveau, P., Trement, F., Walsh, K. & Barker, G. (eds) *Environmental Reconstructions in Mediterranean Landscape Archaeology*, 181 – 91. Oxford: Oxbow Books.

Lizot, J. 1976. *The Yanomami in the Face of Ethnocide*. Copenhagan: International Work Group for Indigenous Affairs.

Lizot, J. 1989. 'Sobre la guerra: una respuesta a N. A. Chagnon (Science, 1988)', *La Iglesia en Amazonas* 44, 23 – 34.

Lorenz, K. 1978. 'The functional limits of morality', in Caplan, A. L. (ed.) *The Sociobiology Debate: Readings on Ethical and Scientific Issues*, 67 – 75. London: Harper & Row.

Lovejoy, A. O. & Boas, G. 1965. *Primitivism and Related Ideas in Antiquity*. New York: Octagon Books.

Lucke, W. & Frey, O.-H. 1962. *Die Situla in Providence*. Berlin: Römisch-Germanische Forschumgen.

Magnin, F. 1993. Le Pays d'Aix: Le Milieu Physique et son évolution récente' in Coutagne, D. (ed.) *Archéologie d'Entremont au Musée Granet*, 25 – 32. Aix-en-Provence: Association des Amis de Musée Granet.

Mahieu, E. 1998. 'The anthropologie à Entremont', *Documents d'Archéologie Méridionale* 21, 62 – 66.

Martin, D. L. & Frayer, D. W. 1997. *Troubled Times: Violence and Warfare in the Past*. Amsterdam: Gordon & Breach Publishers.

Marty, F. 1999. 'Vaisselle et organisation sociale du village de La Cloche (Les Pennes-Mirabeau, B.-du-Rh.) au Ier siècle avant notre ère', *Documents d'Archéologie Méridionale* 22, 139 – 220.

Marty, F. 2002. 'L'habitat de hauteur du Castellan (Istre, B.-du-Rh.) à l'âge du fer. Étude des collections anciennes et recherches récentes', *Documents d'Archéologie Méridionale* 25, 129 – 69.

Marwick, M. 1970. 'Witchcraft as a Social Strain-gauge', in Marwick, M. (ed.) *Witchcraft and Sorcery*, 280 – 95. Harmondsworth: Penguin.

Maschner, H. D. G. & Reedy-Maschner, K. L. 1997. 'Raid, retreat, defend (repeat): the archaeology and ethnohistory of warfare on the North Pacific Rim', *Journal of Anthropological Archaeology* 17, 19 – 51.

Mead, M. 1968. 'Alternatives to war', in Fried, M., Harris, M. & Murphy, R (eds) *War: The Anthropology of Armed Conflict and Aggression*, 215 – 28. New York: Doubleday.

Megaw, J. V. S. 1970. *Art of the European Iron Age*. New York: Harper & Row.

Megaw, J. V. S. 2003. 'Where have all the warriors gone? Some aspects of stone sculpture from Britain to Bohemia', *Madrider Mitteilungen* 44, 169 – 289.

Mercer, R. J. 1999. 'The origins of warfare in the British Isles', in Carman, J. & Harding, A. (eds) *Ancient Warfare*, 143 – 56. Trowbridge: Sutton Publishing.

McCartney, M. 2006. Finding fear in the Iron Age of southern France', *Journal of Conflict Archaeology* 2, 99 – 118.

McCartney, M. 2008. Évidence pour la guerre dans les règlements d'age de fer de la France méridionale. *Proceedings of SOMA 2005 B.A.R. International Series* 1739, 285 – 96.

McCauley, C. 1990. 'Conference overview', in Haas, J. (ed.) *The Anthropology of War*, 1 – 25. Melksham: Redwood Press Limited.

Mignon, J.-M. 2005. 'Les mausolées de Fourches-Vielles à Orange, Valcluse', in Delestre, X. (ed.) *15 Ans d'Archeologie en Provence-Alpes-Cote-d'Azur*, 142 – 43. Aix-en-Provence, Edisud.

Naroll, R. 1962. 'Floor area and settlement population', *American Antiquity* 27, 587 – 88.

Nash, D. 1978. 'Territory and state formation in central Gaul', in Green, D., Haselgrove, C. & Spriggs, M. (eds) *Social Organisation and Settlement*, 455 – 476. Oxford: British Archaeological Reports International Series 47(ii).

Nash, D. 1985. 'Celtic territorial expansion and the Mediterranean world', in Champion, T. C. & Megaw, J. V. S. (eds) *Settlement and Society*, 45 – 68. Cambridge: Leicester University Press.

Nerzic, C. 1989. *La Sculpture en Gaule Romaine*. Paris: Editions Errance.

Osgood, R. 1998. *Warfare in the Late Bronze Age of Northern Europe*. Oxford: British Archaeological Reports International Series 694.

Osgood, R., Monks, S. & Toms, J. (eds) 2000. *Bronze Age Warfare*. Stroud: Sutton Publishing.

Paillet, J.-L. & Treziny, H. 2000. Le rampart hellénistique et la porte charretière de Glanum, in Chausserie-Laprée (ed.) *Le Temps des Gaulois en Provence*, 189 – 90. Marseille: Musée Ziem.

Papi, R. 2001. 'Le armi: i dischi-corazza', in *Colonna* 2001, 120 – 22 (notices du catalogue).

Pearson, L. 1939. *Early Ionian Historians*. Oxford: The Clarendon Press.

Perrot, R. 1971. Étude Anthropologique d'un ossuaire protohistorique l'Aven Plérimond (Var). Lyon: Documents des Laboratoires de Géologie de la Faculté des Sciences de Lyon.

Piggott, S. 1968. *The Druids*. New York: Praeger.

Proctor, D. 1971. *Hannibal's March in History*. Oxford: Clarendon Press.

Price, B. J. 1978. 'Secondary state formation: an explanatory model', in Cohen, R. & Service, E. (eds) *Origins of the State: The Anthropology of Political Evolution*, 161 – 86. Philadelphia: Institute for the Study of Human Issues.

Price, B. J. 1984. 'Competition, productive intensification and ranked society: speculation from evolutionary theory', in Ferguson, R. B. (ed.) *Warfare, Culture and Environment*, 209 – 40. London: Academic Press Inc.

Py, M. 1990. Culture, Économie et Société Protohistoriques dans la Région Nimoise. Rome: Collection de l'Ecole Française de Rome.

Py, M. 1993. *Les Gauloise du Midi: De la Fin de l'Âge du Bronze à la Conquête Romaine*. Aix-en-Provence: Hachette.

Py, M. & Buxó i Capdevila, R. 2001. 'La viticulture en Gaule a l'âge du fer', *Gallia* 58, 29 – 43.

Randsborg, K. 1999. 'Into the Iron Age: a discourse on war and society', in Carman, J. & Harding, A. (eds) *Ancient Warfare*, 191 – 202. Trowbridge: Sutton Publishing.

Rankin, H. D. 1987. *Celts and the Classical World*. London: Croom Helm.

Rankin, H. D. 1996. 'The Celts through classical eyes', in Green, M. J. (ed.) *The Celtic World*. London: Routledge.

Rapin, A. 1999. 'L'armements celtique en Europe: chronologie de son evolution technologique du Ve au Ier s. av. J.-C.', *Gladius* XIX, 33 – 67.

Rapin, A. 2000. 'La statuaire de Roquepertuse, la statuaire d'Entremont', in Chausserie-Laprée (ed.) *Le Temps des Gaulois en Provence*, 81 – 83. Marseille: Musée Ziem.

Rapin, A. 2003. 'De Roquepertuse à Entremont, la grand sculpture du midi de la Gaule', *Madrider Mitteilungen* 44, 223 – 48.

Rapin, A. 2004. 'Pour une nouvelle lecture de la sculpture préromaine de Gaule meridionale', *Documents d'Archéologie Méridionale* 27, 13 – 22.

Robarchek, C. 1990. 'Motivations and material causes: on the explanation of conflict and war', in Haas, J. (ed.) *The Anthropology of War*, 56 – 76. Melksham: Redwood Press Limited.

Rolland, H. 1968. 'Nouvelles fouilles du sanctuaire des Glaniques', *RELig* II, 7 – 34.

Ross, A. 1986. *The Pagan Celts*. London: Batsford.

Roth-Congès, A. 2004. 'La contexte archeologique de la statuaire de Glanon (Saint-Remy-de-Provence, Bouches-du-Rhône)', *Documents d'Archéologie Méridionale* 27, 23 – 43.

Salviat, F. 1989. 'Sculptures de pierre dans le Midi', in Goudineau, C. & Guilane, J. (eds) De Lascaux Grand Lovre. Archéologie et Histoire en France, 498 – 505. Paris: Errance.

Salviat, F. 1990. *Glanum et le Antiques*. Aix-en-Provence: Imprimerie Nationale Éditions.

Salviat, F. 1993. 'La sculpture d'Entremont', in Coutagne, D. (ed.), *Archéologie d'Entremont du musée Granet*. Aix-en-Provence: Association des Amis de Musée Granet.

Séjalon, P. 1998. 'Un atelier de potiers gaulois des années 150/50 av. n. è. à Bouriège (Aude)', *Revue Archéologique de Narbonnaise* 31, 1 – 11.

Sharples, N. M. 1991. 'Warfare in the Iron Age of Wessex', *Scottish Archaeological Review* 8, 79 – 89.

Sim-Williams, P. 1998. 'Celtomania and Celtoscepticism', *Cambrian Mediaeval Celtic Studies* 36, 1 – 35.

Spindler, G. & Spindler, L. 1983. Foreward in *Yąnomamö: The Fierce People* (3rd ed.). New York: Holt, Rinehart & Winston.

Stewart, P. J. & Strathern, A. 2002. *Violence, Theory and Ethnography*. London: Continuum.

Taylor, T. 1992. 'The eastern origins of the Gundestrup Cauldron', *Scientific American* 266, 66 – 71.

Taylor, T. 2001. 'Believing the ancients: quantitative and qualitative dimensions of slavery and slave trade in later prehistoric Eurasia', *World Archaeology* 33, 27 – 43.

Thompson, I. B. 1975. *The Lower Rhône and Marseille*. Oxford: Oxford University Press.

Thorpe, I. J. N. 2003. 'Anthropology, archaeology and the origin of warfare', *World Archaeology* 35, 145 – 165.

Tierney, J. J. 1960. 'The Celtic ethnography of Poseidonius', *Proceedings of the Royal Irish Academy* 60, 189 – 275.

Tiger, L. & Fox, R. 1978. 'The human biogram', in Caplan, A. L. (ed.) *The Sociobiology Debate: Readings on Ethical and Scientific Issues*, 57 – 66. London: Harper & Row.

Tinbergen, N. 1978. 'On war and peace in animals and man', in Caplan, A. L. (ed.) *The Sociobiology Debate: Readings on Ethical and Scientific Issues*, 76 – 99. London: Harper & Row.

Treherne, P. 1995. The warrior's beauty: the masculine body and self-identity in Bronze Age Europe. *Journal of European Archaeology* 3, 1, 105 – 44.

Trément, F. 1999. 'The integration of archaeological, historical and palaeoenvironmental data at the regional scale: the Étang de Berre, southern France' in Leveau, P., Trément, F., Walsh, K. & Barker, G. (eds) *Environmental Reconstruction in Mediterranean Landscape Archaeology*, 193 – 205. Oxford: Oxbow Books.

Turney-High, H. 1949. *Primitive War: Its Practice and Concepts*. Columbia: University of South Carolina Press 1971.

Vandkilde, H. 2003. 'Commemorative tales: archaeological responses to modern myth, politics and war', *World Archaeology* 35, 126 – 44.

Verdin, F. 2002. 'Les Salyens, Les cavares et les villes du Rhône', in Garcia, D. & Verdin, F. (eds) *Territoires Celtiques: Espaces Ethniques et Territoires de Agglomérations Protohistoriques d'Europe Occidentale*, 139 – 48. Martigues: Editions Errance.

Walbank, F. W. 1972. *Polybius. Sather Classical Lectures, v. 42*. Berkeley: University of California Press.

Walsh, K. & Barker, G. *Environmental Reconstructions in Mediterranean Landscape Archaeology*. Oxford: Oxbow Books.

Washburn, W. 1988. *Handbook of North American Indians. Vol. 4, History of Indian-White Relations*. Washington: Smithsonian Institution Press.

Webster, J. 1997. 'Necessary comparisons: a post-colonial approach to religious syncretism in the Roman provinces', *World Archaeology* 28, 324 – 38.

Wells, P. S. 1984. *Farms, Villages and Cities: Commerce and Urban Origins in Late Prehistoric Europe*. London: Cornell University Press.

Wells, P. S. 2002. 'The Iron Age', in Milisauskas, S. (ed.) *European Prehistory: A Survey*, 335 – 83. London: Plenum Publishers.

Wheeler, R. E. M. 1958. Review of 'Entremont': Capitale Celto-Ligure des Salyens de Provence', *The Journal of Roman Studies* 48, 211 – 12.

Whitehead, N. 1990. 'The snake warriors – sons of the tiger's teeth: a descriptive analysis of Carib warfare ca. 1500 – 1820', in Haas, J. (ed.) *The Anthropology of War*, 146 – 70. Melksham: Redwood Press Limited.

Willaume, M. 1993. 'Les objets de la vie quotidienne', in Coutagne, D. (ed.) *Archéologie d'Entremont au Musée Granet*, 107 – 42. Aix-en-Provence: Association des Amis de Musée Granet.

Wilson, E. O. 1978. 'Academic vigilantism and the political significance of sociobiology', in Caplan, A. L. (ed.) *The Sociobiology Debate: Readings on Ethical and Scientific Issues*, 303 – 91. London: Harper & Row.

Woolf, G. 1995. 'Beyond Romans and natives', *World Archaeology* 28, 339 – 50.

Wynne-Edwards, V. C. 1978. 'Intergroup selection in the evolution of social systems', in Caplan, A. L. (ed.) *The Sociobiology Debate: Readings on Ethical and Scientific Issues*, 181 – 190. London: Harper & Row.

THE ANCIENT TEXTS

Appian & White, H. 1979. *The Civil Wars, Books 3.27 – 5*. The Loeb Classical Library. Cambridge, Mass: Harvard University Press.

Aristotle & Kraut, R. 1997. *Politics. Books VII & VIII*. Oxford: Clarendon Aristotle Series. Clarendon Press.

Aristotle & Tredennick, H., Barnes, J. & Thomson, J. A. K. 2003. *The Nicomachean Ethics*. London: Penguin Classics.

Athenaeus & Gulick, C. B. 2002. *The Deipnosopists, Vol. II, Books 3 – 5*. London: W. Heinemann. The Loeb Classical Library.

Caesar & Edwards, H. J. 1917. *The Gallic War*. London: W. Heinemann. The Loeb Classical Library.

Diodorus Siculus & Oldfather, C. H. 1954. *Library of History, Vol. VI, Books 14 – 15*. London: W. Heinemann. The Loeb Classical Library.

Herodotus & Marincola, J. M. & Selincourt, A. 2003. *The Histories*. London: Penguin Classics.

Trogus & Justinus, M. J. 1994. *Epitome of the Philippic History of Pompeius Trogus*. Atlanta, Ga: Classical Resources Series, No. 3. Scholars Press.

Livy & Foster, B. O. 1924. *History of Rome*, Vol. III, Books 5 – 7. London: W. Heinemann. The Loeb Classical Library.

Livy & Foster, B. O. 1924. *History of Rome*, Vol. V, Books 21 – 22. London: W. Heinemann. The Loeb Classical Library.

Lucan & Duff, J. D. 1928. *Lucan (Pharsalia)*. London: W. Heinemann. The Loeb Classical Library.

Nicander & Gow, A. S. F. & Scholfield, A. F. 1979. *The poems and poetical fragments (Greek texts and commentaries)*. New York: Arno Press.

Plato & Bury, R. G. 1926. *Laws, Books 1 – 6*. Cambridge, Mass: The Loeb Classical Library, Harvard University Press.

Pliny & Rackham, H. 1938. *Natural History*. Cambridge, Mass: The Loeb Classical Library, Harvard University Press.

P. Mucius Scaevola. *Chronica Regia Coloniensis (annales Maximi Colonienses)*. 1978. Monumenta Germaniae Historica. Scriptores rerum Germanicarum in usum scholarum separtim editi. Cited in Collis, J. 2003. *The Celts: Origins, Myths and Inventions*. Stroud: Tempus Publishing Ltd.

Polybius & Paton, W. R. 1922. The Histories, Vol. II, Books 3 – 4. London: W. Heinemann. The Loeb Classical Library.

Rufius Festus Avienus & Murphy, J. P. 1977. *Ora Maritima*. Chicago: Ares Publishers.

Seneca & Gummere, R. M. 1920. *Epistles 66 – 92*. Cambridge, Mass: The Loeb Classical Library, Harvard University Press.

Silius Italicus, T. C. & Duff, J. D. 1934. *Punica*. London: W. Heinemann. The Loeb Classical Library.

Stephanus. *Europa*. Fragments cited in Koch, J. T. & Carey, J. 1997. *The Celtic Heroic Age: Literary Sources for Ancient Celtic Europe and Early Ireland and Wales*. Cardiff: University of Wales, Centre for Advanced Welsh & Celtic Studies.

Strabo, Jones, H. L. & Sterrett, J. R. S. 1917. *The Geography of Strabo*. London: W. Heinemann. The Loeb Classical Library.

Timaeus cited in Rankin, D. *Celts in the Classical World*. London: Routledge.

Thucydides, Warner, R. & Finley, M. I. 1972. *History of the Peloponnesian War*. Harmondsworth: The Penguin Classics.